Third Edition

SPICE: A GUIDE TO CIRCUIT SIMULATION AND ANALYSIS USING PSPICE®

PAUL W. TUINENGA
MicroSim Corporation

PRENTICE HALL
Englewood Cliffs, New Jersey 07632

Library of Congress Cataloging-in-Publication Data

Tuinenga, Paul W.
   SPICE: a guide to circuit simulation and analysis using PSpice /
Paul W. Tuinenga -- 3rd ed.
     p. cm.
  Includes bibliographical references and index.
  ISBN: 0-13-158775-7
  1. SPICE (Computer file). 2. PSpice. 3. Electric circuit anlaysis--
Data processing.   I. Title.
TK454.T85 1995
621.3815'01'1353--dc20   94-42313
                              CIP

Editorial director: **DAVE RICCARDI**
Managing editor: **LINDA BEHRENS**
Publisher: **ALAN APT**
Production editor: **JENNIFER WENZEL**
Cover designer: **WENDY ALLING JUDY**
Maufacturing buyer: **LORI BULWIN**
Editorial assistant: **SHIRLEY MCGUIRE**

©1995, 1992, 1988 by Prentice-Hall, Inc.
A Simon & Schuster Company
Englewood Cliffs, New Jersey 07632

The author and publisher of this book have used their best efforts in preparing this book. These efforts include the development, research, and testing of the theories and programs to determine their effectiveness. The author and publisher make no warranty of any kind, expressed or implied, with regard to these programs or the documentation contained in this book. The author and publisher shall not be liable in any event for incidental or consequential damages in connection with, or arising out of, the furnishing, performance, or use of these programs.

TRADEMARK INFORMATION
IBM® PC is a registered trademark of
   International Business Machine Corporation.
PSpice® is a registered trademark of MicroSim
   Corporation.

Printed in the United States of America

10 9 8 7 6 5 4 3 2 1

ISBN 0-13-158775-7

Prentice-Hall International (UK) Limited, London
Prentice-Hall of Australia Pty. Limited, Sydney
Prentice-Hall Canada Inc., Toronto
Prentice-Hall Hispanoamericana, S.A., Mexico
Prentice-Hall of India Private Limited, New Delhi
Prentice-Hall of Japan, Inc., Tokyo
Simon & Schuster Asia Pte. Ltd., Singapore
Editora Prentice-Hall do Brasil, Ltda., Rio de Janeiro

To Louie, Claire, Bill, and Jim
— and our new life.

THE DESIGN CENTER™

Free Software

The **Design Center** includes packages containing schematic capture, simulation with our **PSpice** native mixed analog/digital simulator, and graphical waveform analysis of analog and digital circuit designs. Class instructors can receive complimentary evaluation versions for *both* the IBM-PC and Macintosh by submitting a request on company or educational letterhead to:

Product Marketing Department
MicroSim Corporation
20 Fairbanks
Irvine, CA 92718

Duplication of the diskettes for your students is encouraged.

MicroSim Corporation
The Standard for Circuit Design

CONTENTS

Preface to the Third Edition

SPICE, from the University of California, at Berkeley, is the *de facto* world standard for analog circuit simulation. PSpice®, from MicroSim Corporation, is one of the many commercial derivatives of U. C. Berkeley SPICE. PSpice was the first SPICE-derived circuit simulator available on the IBM personal computer, and was introduced in January 1984 when the IBM-PC was only 29 months old. At this writing, the worldwide installed base of personal computers is estimated at over 150 million units. PSpice is beginning its second decade with nearly 20,000 professional versions, and over 100,000 student versions, in use.

SPICE has proven to be one of the "durables" of our computer age. We should be thankful to the people who created it and the research process that made it possible.

About This Book

Early on, as PSpice became popular, its users found that there are no "how to" references for using SPICE, or its derivatives, the way there are references for database, spreadsheet, and word processing programs. This book's goal is to help users of PSpice, and many other SPICE-derived simulators, understand and use the features of the simulator for their work. Many of these features are only hinted at in other references, notes, or advice from other users, which are the traditional means of help you get for SPICE. This book also tries to be a "friendly alternative" to the terse documentation you receive with most circuit simulation software.

Beyond the syntax and semantics of the SPICE-standard input file, this book also demonstrates the use of PSpice for electrical engineering applications. There is a lot you can do with a circuit simulator that corresponds to what you might do on the lab bench, as well as simulated measurements that go beyond what is possible with lab equipment.

As you might have guessed, from the size of this book (compared to more general books on electronics), some background is assumed. In particular:

- You may be a student or other beginner, but this book assumes that you have a passing acquaintance with electrical or electronic circuits. This may come from formal study of basic electronic components and network analysis. (Do you know what resistors and capacitors are, and how they react to electrical stimulation? Do you recall Kirchhoff's laws?) Or perhaps you picked up a working knowledge from your job, or as a hobby. If you have a circuit whose operation you basically understand, and you want to simulate this circuit to check the details of operation, you probably know enough to understand PSpice.

- This book also assumes that you are able to operate the computer that will run the simulator, as well as create the input file for the simulator. Perhaps a friend, or another kind person, will help you with this. Also, you may want to consult the PSpice *Circuit Analysis User's Guide*, from MicroSim Corporation, for the details of how to get started in your environment.

Circuit designers actively using SPICE and PSpice found that the first and second editions provided many useful insights and explanations that helped in their work and use of circuit simulation. Throughout this text, I have tried to offer material that you will not see in any other book.

Complete, real-life examples are **not** featured in this text. Regretfully, there are too many of these (and hundreds of thousands of engineers creating new ones). This is not a book about electronic circuit design, but a book about the basics and techniques for simulating these circuits. Let me illustrate this with a true story. In college, my father studied electrical engineering with a classmate that had been putting off fixing the toaster at home. His mother assumed that her son, preparing to be an electrical engineer, should be able to fix a toaster — so, she explained to her friends, "I guess he just hasn't gotten to the chapter on toasters, yet." You see, she thought learning electronics was like learning to spell; just memorize all of the rules and examples. Electronics is not "learned by example" only, and neither is circuit simulation. However, I did include an example circuit in the appendix to illustrate how circuit simulation helps you in improving the design of a circuit.

How This Book is Organized

Analog circuit simulation, as a topic, tends to be self-referential, making it difficult to find a starting point and a natural path to visit important topics. Also, covering each topic "from soup to nuts" may be overwhelming for the first-time student, but useful for the practitioner. So a compromise must be made... some readers will like it, and some won't.

I chose to save the area of device modeling for last consideration, since this is practically an area unto itself and independent of the workings of the simulator. Most of the text uses examples containing only sources and linear devices. The most difficult analysis, transient analysis, becomes the end of our path; DC-bias analysis becomes the starting point, since, with non-linear elements, the operating point would be needed to perform a small-signal analysis. So the progression of DC, AC, and transient, are the "guiding stars" for navigating the subject of SPICE, with each chapter building on the material in the previous chapter(s).

I regret that this book cannot be made to suit every electronics course. Each professor and lecturer has his/her own agenda for teaching electronics: Prentice Hall and I have received many conflicting suggestions for improving on the first edition.

We expect, and welcome, comments on this edition and will strive to work these into future editions of the compromise that is this book.

WHAT'S NEW IN THE THIRD EDITION

Two new chapters have been inserted in this edition:

- Chapter 14 explains the use of "Monte Carlo methods" in PSpice for statistically computing estimates of how circuits will behave with variations in component values (for example, what fraction of circuits will work within specification). This technique is general purpose, working for any circuit, and provides results without making approximations. An example is given for two types of filter circuit and how they compare when analyzed this way.

- Chapter 15 explains the use of sensitivity calculations and "worst-case analysis" in PSpice for discovering the maximum range of circuit performance and the causes of extreme operation. These techniques are used to identify effective changes to improve the quality of circuit operation (for example, which components need to have tight tolerance and which can be of lower quality and less expensive). An example is given of how worst case analysis may give false results, which are less extreme than what is possible.

These two chapters are engineering oriented and quite apart from the usual "See Spot. See Spot run." material that shows how you can duplicate, via simulation, what you could calculate or measure at the bench. These chapters show how you can gain insight and extract useful answers for circuit problems that are difficult or impossible to find by traditional methods.

The material from Chapter 14 (second edition) has been moved to Chapter 16. Also, many clarifications and small additions have been made to the material from previous editions.

ACKNOWLEDGMENTS

The unwavering support of many people made this book possible. Foremost, I thank my family for their patience while I combined the effort on this book with my other work, which kept me away from them. Also, I thank the people at MicroSim and my editor at Prentice Hall, Alan Apt, for their help and support. Finally, I thank the legions of PSpice users whose many questions and comments provided the basis for this book.

Paul W. Tuinenga
Folsom, California

ADDITIONAL ITEMS AVAILABLE

Prentice Hall makes this book and the Student Version of PSpice available in several formats. When ordering either the book, or one of our book/disk packages, please refer to the title code number (below) on your order:

1. "SPICE: A Guide to Circuit Simulation and Analysis Using PSpice" Third Edition by Paul Tuinenga (book only) #15877-4

2. Book/Disk package with IBM-PC 5¼" DS/DD diskettes #43610-5

3. Book/Disk package with IBM-PC 3½" HD diskettes #43604-8

The professional version of PSpice is available through MicroSim Corporation and its distributors. PSpice is available for workstations and minicomputers in addition to the PC-based versions. Please check directly with MicroSim for product availability by calling (714) 770-3022.

PSpice helps you simulate your electrical circuit designs **before** you build them. This lets you decide if you need to make changes, without touching any hardware. PSpice also helps you check your design **after** you think it is complete. This lets you decide if the circuit will work correctly outside your office, in the real world, with good production yield. In short, PSpice is a simulated "lab bench" on which you create test circuits and make measurements. However, PSpice **will not** design the circuit for you.

The practical way to check an electrical circuit design is to build it. However, by the early 1970s, the components that were connected on an integrated circuit had become much smaller than individual discrete components. Physical effects that were negligible for normal circuits, such as a stereo amplifier, became important for these microcircuits. So the circuits could not be assembled from components in the lab and give the correct test results — the circuit had to be either (i) physically built, which is expensive and time consuming, or (ii) carefully simulated using a computer program. This is why the acronym SPICE stands for $\underline{S}$IMULATION $\underline{P}$ROGRAM WITH $\underline{I}$NTEGRATED $\underline{C}$IRCUIT $\underline{E}$MPHASIS.

WHAT IS PSPICE?

PSpice is a member of the SPICE "family" of circuit simulators, all of which derive from the SPICE2 circuit simulator developed at the University of California, Berkeley, during the mid-1970s. SPICE2 evolved from the original SPICE program, which, as it turns out, evolved from another simulator called CANCER that was developed in the early 1970s. Tremendous effort over this relatively short time created a simulator with algorithms that are robust, powerful, and general; SPICE2 quickly became an industry standard tool. Since this development was supported using U.S. public funds, the software is "in the public domain," which means it may be freely used by U.S. citizens. The software is improved by U.C. Berkeley to the extent that it supports further research work. For example, SPICE3 is a redesigned implementation of the SPICE2 program that fits into U.C. Berkeley's computer-aided design (CAD) research program. SPICE3 is not better than SPICE2 in the way that SPICE2 was an advance over the original SPICE program; rather, it is designed to be a module in the U.C. Berkeley CAD research effort. Neither SPICE2 nor SPICE3 is supported by U.C. Berkeley like commercial software, nor does U.C. Berkeley provide consulting services for these programs. This lack of support led to commercial versions of SPICE that have the kind of support industrial customers require. Also, many companies have an in-house version of SPICE that has modifications to suit particular needs.

PSpice, which uses the same major algorithms as SPICE2 and is compatible with its input syntax, shares this emphasis on microcircuit technology. However, the electrical concepts are general and are useful for all sizes of circuits (for example, power generation grids) and a wide range of applications. For instance, the simulator has no concept of large or small circuits; microvolts or megavolts are "just numbers" to PSpice. As long as PSpice is able to solve your circuit matrix, it will do so. This makes PSpice "technology independent" and generally useful. On the other hand, a simulator makes no assumptions about how the circuit should behave; for example, PSpice is not concerned that 0.03-watts output power does not make for a very loud audio amplifier. You have to look at the results to see if they make sense in your application.

For discrete circuits (circuits made of individual parts assembled on a circuit board) PSpice has a variety of uses. Like the integrated circuit designs mentioned before, your designs are pressed for schedule time, budget expense, and manufacturing yield. PSpice can offset these constraints by enabling you to

- Check a circuit idea before building a breadboard (even before ordering the parts). Simulated results are free of crosstalk, voltage and temperature instabilities, grounding restrictions, and other prototyping headaches.

- Try out ideal, or "blue sky," operation by using ideal components to isolate limiting effects in your design.

- Make simulated test measurements that are:
    · difficult (due to electrical noise or circuit loading),
    · inconvenient (special test equipment is unavailable), or
    · unwise (the test circuit would destroy itself).

- Simulate a circuit many times with component variations to check what percentage will pass "final test," and to find which combinations give the "worst-case" results.

Once you become familiar with PSpice, you will find that it can substitute for most (but not all) of your breadboard work. Like any new tool, experience is required to get the most benefit from it.

The PSpice control statements are easy to learn and use. These statements, which are collected in a file called the "circuit file"[†] to be read by the simulator, are usually self-contained and may be understood without referring to any other statements. Moreover, each statement has so little interaction with other statements that it has the same meaning regardless of context. So the language of PSpice is easy to learn because you can focus on each statement type, master it, then move on to another. Also, you will see that you do not need to know many statement types to

---

[†] Many users think of the file as a "program" in the sense that PSpice interprets the file to perform the simulation, much the way a BASIC interpreter runs a BASIC program.

get started. Most of your difficulty will probably come from learning and operating your computer system.

OTHER SPICE-BASED PROGRAMS

The commercially supported versions of SPICE2 fall into three groups:

- The original group of mainframe-based versions, including HSPICE from Meta-Software and I-SPICE from NCSS timesharing. HSPICE focuses on the needs of the integrated circuit designer with special device model support. I-SPICE focuses on "interactive" circuit simulation and graphics output (an innovation for mainframe users).

- The IBM-PC based programs (other than PSpice), including IS-SPICE from Intusoft. These are often direct adaptations of SPICE2 or SPICE3 and in most cases no changes have been made — even to correct errors. Without serious support for the simulator, these programs fall into a more "hobbyist" class of product. However, interesting additions include pre-processors or shell programs to manage input and provide interactive control, as well as post-processors for refining the normal SPICE output.

- Advanced programs, with "innards" that are significantly overhauled, or are entirely new, but which use the same direct-method simulation algorithms as U. C. Berkeley SPICE, including AccuSim from Mentor Graphics, PSpice from MicroSim, Spectre from Cadence Design, and ViewSpice from ViewLogic Systems. Many additions and improvements are available from these products. Also, these advanced simulators have options to extend simulation capabilities and interpret results.

The "growth" portion of the analog circuit simulation business is the last group, especially with the rapidly expanding market for engineering workstations. While U.C. Berkeley remains at the forefront of computer-aided tools for engineering, as a practical matter the complete simulation products come from industry. Most, if not all, of the techniques you will see in this text are applicable to these products.

HOW THIS BOOK WORKS

This book adopts a "graduated example" or "tutorial" approach to learning about circuit simulation. It is tempting to load new software and try some examples, well before reading the instructions, so we will attempt to channel this urge toward learning about the simulator. We will start by building a simple circuit, making some DC "measurements," and then we will move on. Often, the biggest hurdle is running a simulator the first few times. After that you start to focus on the electronics you are simulating and how best to measure what you want to discover.

The details of the semiconductor models are described later. These models are independent of the methods for using the simulator. We will do without them entirely in most of the examples.

An abridged summary of the control statements and device descriptions are at the end of this book. These items appear as appendices because this information is not the *raison d'être* of the book, and is applicable to the SPICE-like simulators in only a general sense. You should try to obtain a detailed guide to the simulator you will be using.

Not "everything you ever wanted to know about SPICE..." is in this book. Missing are some topics that I had hoped to include, as well as additional depth of coverage of the topics that do appear. These fell victim to the schedule for completing the text. Additional topics, without doubt, will become obvious from reader comments. Perhaps in another edition...

CHAPTER 1

GETTING STARTED

Let us begin with a quick circuit to introduce you to running PSpice. This will show you the basics of a circuit simulation without the complications of rules, details, exceptions, and so on, and quickly get to a successful result. Later, we will explore the wide range of features and ways of combining these to express complex circuit functions.

Sometimes the examples will intentionally omit features of the simulator to concentrate on a particular topic. These features are necessary for normal use, and we will get to them in due course. The examples are brief to demonstrate an idea, and there is the danger that, in not explaining somewhat unrelated items, they may mislead you. If this happens it was not intended, but is just a problem with this approach.

Sometimes the examples will repeat some of what was done already.

1.1 A SMALL CIRCUIT

The best way to learn a circuit simulator is to "do" simulations. Running any simulation requires several basic accomplishments. It requires that you

- create the input file, or "circuit file,"
- run the simulator, without error(s),
- find where the output went, and
- inspect the output.

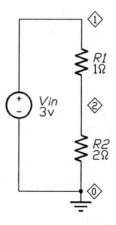

FIGURE 1-1   Schematic for small-circuit example.

Naturally, you would start with a small circuit that you know, by inspection, will work. We will use the circuit shown in Figure 1-1 for starting out. In PSpice, the circuit file to simulate this circuit is

```
* Resistor divider circuit
VIN 1 0 3.0volt
R1  1 2 1.0ohm
R2  2 0 2.0ohm
.END
```

How you run this simulation will depend on the system you are using. You will need to learn to use a text editor to create the input file. Then you will run PSpice, specifying the input file you created. If everything works, PSpice will read your input file called (for example) TEST.CIR and place the results in an output file called TEST.OUT. The same text editor you used for creating the input file can also be used to inspect the output file. This output file may also be directed, by you, to a printer.

1.2 THE INPUT FILE

Now to describe, and explain, the circuit file. PSpice always expects the first line of the circuit file to be a title line. You can leave it blank, but circuit description cannot start until the second line of the file. The examples in this book will sometimes start the title line with * (which also indicates a comment line) by force of habit on my part. This is not necessary. What is necessary is the last line, .END (you **must** include the "."), which completes the description of the entire circuit including any simulation controls. You use .END because PSpice will let you start another, completely different, circuit simulation right after .END. Between the first and last line, the circuit file may be in almost any order.

All of the circuit elements, or devices, in the circuit file are connected (in the sense that you would solder their leads together) by circuit nodes. You may think of these nodes as the connecting wires, or lines, in a circuit schematic. In SPICE2 these nodes are positive integers, including 0 (zero), which is reserved to mean "ground." PSpice **does not require** that you use integers (any text string will do) but 0 is still "ground." Every circuit file must have a ground node, as a reference, and every other node in the circuit file must have a DC current path to ground. This is one of the requirements of the SPICE algorithms.

Along with requiring a ground node, PSpice also requires that all terminals be connected to at least one other terminal. This is a precaution against dangling wires. Even though you may do this on the lab-bench, it is considered an error by the simulator.

The circuit file for our example uses only two-terminal devices — a voltage source and the resistors. A separate line is used to describe each element in the circuit. The basic syntax is

   name node node [[node...]] value

TABLE 1-1 Forms used to describe PSpice statements.

Form	Meaning
ELEMENT	Monospaced type indicates a keyword that you **must** use. This type is also used in the text for examples and fragments of PSpice input and output files.
placeholder	Italic type indicates an item that you **must** supply, such as a name or a value. Italics are also used in the text to highlight significant words.
[[*option*]]	Items inside double square-brackets are **optional**.
repeat...	A horizontal ellipsis (...) following an item indicates that more items having the same form may appear.
{this|that}	Curly brackets and a vertical bar indicate a choice between two or more items. You **must** choose one of the items.

These forms may be combined: {X|Y}*name* indicates a *name* that **must** start with an X or a Y; [[*node...*]] indicates an **optional** list of *nodes*.

where the syntax forms used in the text are described in Table 1-1.

Note that there are no one-terminal devices in PSpice. Devices with more than two terminals use basically the same form, but with more optional [[*node*]] items. The device *value* is a number, either decimal or floating point, that describes the size of the device. You will see later that there are a variety of ways to express the same value, including a metric suffix. After the value you may include a unit, such as volt or ohm, for your own use; PSpice actually ignores these (to the extent that they aren't confused with one of the metric suffixes).

Any line may be a comment line by starting it with "*" in the first column. This allows you to document your circuit file for others, unfamiliar with the circuit file, or for yourself when, after some time, you too will be unfamiliar with the circuit file. Blank lines are ignored; use them to separate sections of your circuit file.

PSpice also allows you to insert comments on any line by starting the comment with a semicolon. Everything on the line after the semicolon is ignored; for example:

```
Rbias 2 3 45 ; this is the biasing device and had better not fail!
```

EXERCISE 1.2-1

Create and run this simulation on your system. Look at the output file. Read the next section.

1.3 THE OUTPUT FILE

Take a look at the output file of our example. The output file is divided into sections, like an engineering report about the analysis of the circuit. Each section is separated by a banner displaying the date and time, and the version of PSpice used. The first section is a copy of the input file, which, like a restatement of an engineering problem, will remind you about the simulation. For example:

```
**** 08/01/94 01:15:35 ******** PSpice 6.1 (July 1994) ***********

* Resistor divider circuit

****      CIRCUIT DESCRIPTION

*******************************************************************

VIN 1 0 3.0volt
R1  1 2 1.0ohm
R2  2 0 2.0ohm
.END
```

As it turns out, this example did not specify any type of simulation, such as frequency response; however, PSpice assumes that at least you wanted a DC bias-point to be calculated. This is a calculation of the voltages the nodes would have if the circuit were quiescent. These are printed in the next section of your output file:

```
**** 08/01/94 01:15:35 ******** PSpice 6.1 (July 1994) ***********

* Resistor divider circuit

****      SMALL SIGNAL BIAS SOLUTION      TEMPERATURE = 27.000 DEG C

*******************************************************************

 NODE  VOLTAGE     NODE  VOLTAGE     NODE  VOLTAGE     NODE  VOLTAGE

(   1)  3.0000  (   2)   2.0000
```

Calculating the node voltages means the currents through the devices are calculated. PSpice checks all the devices that supply current to the circuit and totals the quiescent power dissipated by the circuit. These values are printed next:

```
VOLTAGE SOURCE CURRENTS
NAME           CURRENT

 VIN       -1.000E+00

TOTAL POWER DISSIPATION   3.00E+00 WATTS
```

Finally, PSpice provides details about the computer aspects of the simulation, which are generally called "job statistics." We did not ask, in the input file, for detailed statistics. So all we will see in the output file is the time spent on the simulation by the computer, in seconds:

```
JOB CONCLUDED

TOTAL JOB TIME           .06
```

To a certain extent, you may control the amount and detail of the information in the output file. This is done through the .OPTIONS[†] statement. For instance, you may

---

[†] The .OPTIONS statement, and other statements, are described in Appendix B.

not want a copy of the input file in your output file. Or, as mentioned before, you may want a detailed report of the job statistics.

EXERCISE 1.3-1

Using the small circuit example, did the output file show (correctly) node 2's voltage as 2 volts? Experiment by changing the circuit file — leave out, or add, something — and see what errors PSpice will check.

EXERCISE 1.3-2

Modify the previous exercise's circuit file by swapping any of the lines, except the first or last line. See if PSpice gives different results. Try adding some comment lines and blank lines to the circuit file.

1.4 COMPONENT VALUES

All quantities, or values, in PSpice may be expressed as decimal or floating-point values traditionally used by computer programs. The decimal numbers should be familiar; for example:

```
3.14   -13.7   .0045
```

Floating point values scale a decimal number by a power of ten, where the letter "E" (for "exponent") separates the decimal number from the start of the integer exponent, so that .0045 can be written 4.5E-3, which means 4.5×10^{-3}.

Older SPICE versions also allow you to use "D" instead of "E." This is a holdover from the FORTRAN programming language, where "D" meant that the number was stored with greater precision ("double precision"). PSpice will also accept the "D" format, but the storage precision is selected depending on the needs of the simulator.

Also, PSpice lets you use a metric-like suffix to express a value. These suffixes multiply the number they follow by a power of ten, except for MIL. Using the suffix notation allows values written into the circuit file to look like the values on a circuit schematic. This is a great convenience that removes a source for most simulation errors: using the wrong component values.

The following are the power-of-ten suffix letters, used by PSpice, along with the metric prefixes and scale factors they represent:

F	*femto-*	10^{-15}
P	*pico-*	10^{-12}
N	*nano-*	10^{-9}
U	*micro-*	10^{-6}
M	*milli-*	10^{-3}
K	*kilo-*	10^{+3}
MEG	*mega-*	10^{+6}
G	*giga-*	10^{+9}
T	*tera-*	10^{+12}
MIL	(0.001")	25.4×10^{-6}

where the last suffix, MIL, provides an English-to-metric conversion of integrated circuit device sizes.

Using the exponential and suffix notation lets you express the same value many ways; for example:

        1050000   1.05E6   1.05MEG   1.05E3K   .00105G

all represent the same value to PSpice.

The previous list of suffixes was written in capital letters because the original SPICE simulators allowed only capital letters in the circuit file. One problem with this is the confusion between the standard use of "m" for *milli-* and "M" for *mega-*, which was resolved by requiring the input to be MEG for *mega-*. To maintain compatibility PSpice still requires that you use MEG, or meg, for *mega-*.

Other letters that are not suffixes may be used with a number, but these are ignored by PSpice. Thus you may write

        10   or   10volts   or   10ohms

to make your circuit file easier to understand without changing the meaning of the value. Moreover, once a valid suffix is read by PSpice, the remaining letters are ignored. You may also write

        10pF   or   10picoamps   or   10picoseconds

These will all represent the value 10×10^{-12}.

EXERCISE 1.4-1

Using the small circuit example, try changing all the 0 nodes to 3, so there is no "ground," and see what happens. Then, try disconnecting R1 and R2 by adding a new node to the circuit file and see what errors result.

In the previous chapter PSpice calculated the DC bias-point for the circuits you entered. PSpice must calculate the bias-point before proceeding to any other type of analysis, since it must determine the operating point of the circuit (the voltages at each node and currents through each device). If you were to physically build a circuit and attach a power supply to it, when you start the supply the circuit will bias itself at its DC operating point. For most circuits this is a stable condition, without oscillation, and PSpice will arrive at a DC solution to the circuit. Later, we will cover more stubborn circuits. These circuits work on the lab-bench, but PSpice will need help to calculate the DC bias-point.

While PSpice always calculates a bias-point before proceeding, it will print out the results of this calculation **only** if

- there are no analyses specified, or

- you include a .OP or .AC statement in your circuit file. (The .AC statement is described in §6.2.)

Even if you want only the DC bias information it is helpful to include the .OP statement to remind you later that the DC bias information was what you were after in the circuit. In a sense, .OP is one of the analyses that PSpice will perform on your circuits.

Using only the DC bias-point analysis, we can demonstrate some electrical laws that PSpice follows in calculating voltages and currents. But first we will learn more about the basic components for building circuits.

2.1 PASSIVE DEVICES

The passive devices are resistors, capacitors, and inductors:

- Resistors limit, or resist, the flow of electrical current, following the law $V = I \times R$, where V is the voltage (in *volts*) across the resistor, I is the current (in *amperes*) through the resistor, and R is the resistance value (in *ohms*).

- Capacitors store energy in an electrostatic field, following the law — C where Q is the induced charge (in *coulombs*) on the "plates" or, V is the voltage (in *volts*) impressed on the "plates," and ance value (in *farads*).

7

- Inductors store energy in an electromagnetic field, following the law $\lambda = I \times L$, where λ is the induced magnetic flux (in *Weber-turns*) around the inductor, I is the current (in *amperes*) through the inductor, and L is the self-inductance value (in *henries*).

Fortunately, most of the Rs, Ls, and Cs we use on the lab-bench are nearly ideal and for our purposes we can consider them to be ideal. In PSpice we can specify these devices merely by using the appropriate letter as the first letter of the device name:

R*xx*	for resistor
L*xx*	for inductor
C*xx*	for capacitor

The *xx* represents any other letters or numbers you want to use to finish the name of the device.

To specify the device in the circuit file we include the name of the device, how it is connected into the circuit, and its value. PSpice uses the basic electrical units for voltage (volts) and current (amps) and uses the basic electrical units for device values: ohms, farads, and henries. Here are some example devices:

```
R12   5 2 15K        is a 15-kilohm resistor (15,000 ohm)
C2   12 3 1.8u       is a 1.8-microfarad capacitor (0.0000018 farad)
L3    7 6 10m        is a 10-millihenry inductor (0.01 henry)
```

2.2 COMPONENT NAMES

As you saw above, the names for devices start with a letter reserved for that device. The first letter tells PSpice what kind of device you are about to describe. These letters correspond to the standard ones used on circuit schematic diagrams for labeling devices. For instance, if you used "R17" as the label of a resistor in a schematic, you would probably use R17 in your circuit file as the name of that resistor. The remaining letters of the component name may be alphabetic letters, and numbers. In PSpice, you may also include the underscore "_" and the dollar sign "$" characters. Uppercase or lowercase letters may be used, but PSpice is not sensitive to which case is used, so that

```
RBIAS
Rbias
rbiaS
```

all refer to the same device. The maximum length possible for component names is longer than 80 characters. Practically, the length of the name you use for a component depends on how much typing you want to do.

Older SPICE versions limited component names to eight characters, due to limitations of the computer language they were written in   s the need to conserve memory on the (then) current generation o   when "core" memory referred to the tiny, hand-strung magnetic rin   nly one bit).

Of course, the first character specified the component type, so there were only seven characters left for making the name unique and identifiable.

Some of the statements in SPICE can be detailed and long, especially in PSpice where you are allowed to have long names for devices and nodes. As a convenience you may split a line wherever you could normally use a space character, and continue on the next line. However, the first character on the continuation line must be a plus sign "+" to indicate that it is a continuation line; for example:

```
ResistorWithLongName ConnectedToOneNode AndToAnotherNode
+ 120ohms
```

2.3 INDEPENDENT SOURCES

To simulate your circuits you will also need some way to tell PSpice what is "exciting" or supplying electrical power to the circuit. For this we use independent sources that supply a fixed voltage level or current flow. We specify these sources in a way that is similar to the passive devices described earlier: name, connecting nodes, value. As you might have guessed

Vxxx   is a voltage source, and
Ixxx   is a current source

Remember, PSpice uses the basic electrical units for values so the following examples are easy to understand:

```
VIN 3 0 1.2K          is a 1.2-kilovolt source (1,200 volts)
I4 12 2 15m           is a 15-milliamp source (0.015 amp)
```

A voltage source is like a battery, or lab-bench power supply. Using the positive current convention, current flows out from the positive terminal (first node), through the circuit, and then into the negative terminal (second node). This is the conventional current flow taught to students. But why, in our first example, did PSpice calculate with the supply current as a negative value? Because whenever you ask PSpice to print the value of a current through a device, "through" means into the first terminal and out of the second terminal. In this case the current was flowing out of the positive (first) terminal, so the current has a negative value.

A current source provides a fixed value of current to the circuit. However, its current flows (again using positive current) into the positive terminal (first node), through the source, and then out of the negative terminal (second node). This is the opposite direction of the voltage source, but is consistent with reporting the current through the device. Also, current flows from the more positive potential to the more negative potential — in this case, through the current source device.

EXERCISE 2.3-1

Using the circuit from the first exercise, replace the voltage source VIN with a current source of value 1 amp. How did the output from PSpice change?

2.4   Parameters and Functions

Sections of electronic circuits are often designed through the use of formulas. For example, the unloaded output voltage of a two-resistor divider is determined by a ratio with the sum of the resistance values in the denominator:

$$\text{division ratio} = \frac{R_B}{R_A + R_B}$$

But to use even this simple circuit in a simulation, or in an actual design, you will need to know the component values.

However, in creating a design, you may not want to commit to particular component values, because you have only general constraints for the circuit when you are getting started. PSpice helps you by allowing parameters and formulas as part of the circuit description. For example, say that you want the voltage divider to provide 20% of the input voltage, and load the input by 50,000 ohms, at most, which will then be the combined value of the resistors. Of course, you can probably work this example in your head, but you can also have PSpice calculate the resistance values needed to satisfy the design. The design values you want specified will become parameters, which are defined by the following form:

```
.PARAM name = value ...
```

This form is very similar to many computer programming languages. Note that you may define more than one parameter on the same line. In our example we would probably define the following parameters (remember, you **must** include the "." to start the statement):

```
.param load  = 50K
.param ratio = .2
```

Now, our little divider can be defined simply with the following lines:

```
Ra in out {load*(1-ratio)}
Rb out  0 {load*ratio}
```

where the formulas, contained by the curly brackets "{ }", are evaluated by PSpice. The result is the component value. The formula doesn't need to be complicated, and you may use formulas almost everywhere a numeric value could be used. Later, if you change your mind and decide that the input resistance should be reduced to 10,000 ohms, you only need to change one parameter and the circuit will continue to operate as planned.

This may not seem very useful, since the previous example was very simple (to demonstrate the concept). But let's say we make it a bit more interesting by designing the voltage divider to provide a fixed output resistance:

$$\text{output resistance} = R_A \| R_B = \frac{R_A \cdot R_B}{R_A + R_B}$$

This design constraint would probably mean you need to use a calculator each time you change this part of the design, or you could let PSpice calculate the values:

```
Ra in out {load/ratio}
Rb out  0 {load/(1-ratio)}
```

The formula can be **any** length and span many lines in the file. Besides the standard arithmetic operators you may also use the functions described in Table 2-1.

Another way to build complicated formulas is to use a macro-like expansions in your formulas. These "macros" are defined using the .FUNC statement, which has the following form:

.FUNC *name* (⟦*arg...*⟧) { *body* }

where the use of *name*, with its arguments, will be replaced with the *body* of the .FUNC definition. (Remember, you **must** include the "." to start the statement.) Using the resistor divider example, instead of creating a new .PARAM to simplify a formula, we could have defined a .FUNC to do this job:

```
.func top(load,ratio) {load*(1-ratio)}
Ra in out {top(load,ratio)}
```

Using a .FUNC for user-defined "functions" not only has the advantage of being used in any formula, it can be used more than once and with different arguments.

TABLE 2-1 Functions available for PSpice formulas.

Function	Equivalent	Meaning
ABS(x)	$|x|$	absolute value
SQRT(x)	$x^{1/2}$	square root
EXP(x)	e^x	exponential
LOG(x)	$ln(x)$	logarithm base e
LOG10(x)	$log(x)$	logarithm base 10
PWR(x,y)	$|x|^y$	power
PWRS(x,y)	$sgn(x)|x|^y$	signed power
SIN(x)	$sin(x)$	sine (x in radians)
COS(x)	$cos(x)$	cosine (x in radians)
TAN(x)	$tan(x)$	tangent (x in radians)
ASIN(x)	$sin^{-1}(x)$	arc sin (result in radians)
ACOS(x)	$cos^{-1}(x)$	arc cosine (result in radians)
ATAN(x)	$tan^{-1}(x)$	arc tangent (result in radians)
LIMIT(x,min,max)	*min* if $x < min$ *max* if $x > max$ otherwise x	limit result to *min* and *max*
TABLE($x,x_1,y_1,\cdots$)	$f_{table}(x)$	lookup table, interpolate between points x_i, y_i

Exercise 2.4-1

The attenuation of a low-pass RC filter can be expressed as

$$\text{attenuation (in dB): } A = 10 \cdot log\left(1 + \left(f_x/f_0\right)^2\right)$$

where f_x/f_0 is the ratio of a given frequency to the cutoff frequency. Change the RC filter example to include a specification for an attenuation at a given frequency.

2.5 Ohm's Law

PSpice calculates values according to many laws of physics including Ohm's law, which we have seen already. The output from your first exercise showed the current flowing through the resistors and the voltages across each resistor. Take a moment to check that Ohm's law was followed for each resistor.

2.6 Kirchhoff's Network Laws

Consider the circuit in Figure 2-1, with the equivalent circuit file:

```
Resistor bridge
VIN 1 0 10
R1   1 2 2
R2   1 3 1
R3   2 0 1
R4   3 0 2
R5   3 2 2
.end
```

After you run PSpice on this circuit, add up the voltage drops around any of the loops in the resistor network. For example, the loop of resistors R2, R5, and R1 has voltage drops (going clockwise) of 4 volts, 2 volts, and ±6 volts. The sum of these voltages is always zero, which demonstrates one of Kirchhoff's network laws: *the algebraic sum of the potential drops around any closed loop in a network of conductors is always zero.*

Exercise 2.6-1

Sum the voltage drops around these loops: R3+R5+R4 and R1+R2+R3+R4.

Now we are going to try something a little different. Earlier we saw that PSpice outputs the current through the voltage source. If we use a voltage source with a value of 0 volts, we can insert this into the circuit and measure (as with an ammeter) currents flowing through the circuit. Let us change the circuit to measure some currents, as in Figure 2-2, with the equivalent circuit file:

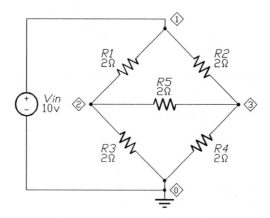

FIGURE 2-1 Schematic for resistor bridge circuit.

```
Resistor Bridge
VIN  1   0   10
R1   1   21  2
V1   21  2   0
R2   1   3   1
V3   22  2   0
R3   22  0   1
R4   3   0   2
R5   3   23  2
V5   23  2   0
.end
```

After you run PSpice on this circuit, add up the currents for V1, V3, and V5. These turn out to be zero, which demonstrates one of Kirchhoff's network laws: *the algebraic sum of the currents coming into any junction in the network is always zero.*

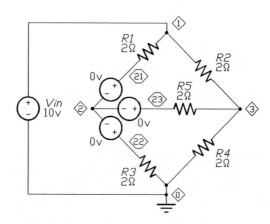

FIGURE 2-2 Resistor bridge circuit, with zero-volt sources.

Exercise 2.6-2

Change the polarity of one of the zero-volt sources and see how the output from PSpice changes. Does Kirchhoff's law still hold? Now, try inserting zero-volt sources around node 3 and check the output from PSpice.

2.7 Capacitors in DC Circuits

Capacitors block sustained, or DC, current. The only time current flows through the capacitor is when the charge is collecting on, or being removed from, the "plates." This means that the voltage across the capacitor is changing, which is not the DC case. Of course in a real circuit, once the power is supplied, there is a transient during which some capacitors charge up to their final values. But the result is the same as if these capacitors did not exist and the connections to each capacitor were left dangling. Inside PSpice, this is exactly how capacitors are treated for DC calculations.

Since the capacitors in PSpice are perfect (that is, without any leakage) it is important that no sections of your circuit become isolated by "ignoring" the capacitors. This means that every node of your circuit needs some path for DC current to ground, however convoluted, so that bias levels may be determined for every node. If you have a circuit with a node that is isolated by perfect capacitors, for example, node 1 in the following circuit fragment:

```
C1   1   0   1pF
C2   1   2   1pF
V1   2   0   3volts
```

there is no way, in theory, to determine the DC level of this node. In a real circuit, leakages in the dielectric of the capacitor would prevent such a node from attaining a "zillion" volts.

PSpice checks for isolated nodes before starting any simulations. If you accidentally isolate a node you will receive an error message similar to

```
Node ... is floating.
```

to indicate this problem. The simulation cannot proceed until this error is corrected.

For the example shown above, you should connect a large-value resistor, say 1-gigaohm, to the isolated node. The other end of the resistor would be connected to ground, or whatever voltage level you wanted to use as the bias level.

Exercise 2.7-1

Insert a capacitor into one of the "legs" of the resistor bridge circuit (see earlier exercises). Run PSpice and note how the bias level changes. Try different "legs" of the bridge and check that the results were as you expected.

EXERCISE 2.7-2

By inserting two capacitors, isolate a resistor in one of the "legs" of the resistor bridge circuit (see earlier exercises). Run PSpice and find the error message resulting from this situation.

2.8 INDUCTORS IN DC CIRCUITS

Inductors, which are essentially a coil of wire, conduct DC current so they do not have the same restrictions as capacitors (previously discussed). However, they have a different restriction (of course, there's always a catch!) because of how inductors are simulated when the analysis includes time, such as AC response or transient simulation. In these cases, the inductor develops a voltage across its winding in response to the changing magnetic flux within the windings. The total voltage developed is the sum of the following:

- flux changes due to current in the winding itself, and

- flux changes due to current in any other winding with a magnetic field coupled to the winding in question.

For these two cases, the ratio of the voltage developed, due to the change in current, is called "self inductance" and "mutual inductance," respectively.

These voltages, which are developed by flux changes, are modeled in PSpice as time-varying voltage sources. So far this is not a problem, except when you try to connect an inductor directly across a voltage source, even if the voltage source's value is zero. This situation is called a "voltage loop," which, as it sounds, means a circular path of voltage sources without any intervening resistance to limit the current to a finite value. To strictly check for all voltage loops, PSpice treats inductors as though they were voltage sources. For the DC case they are "shorts" or zero-volt sources, and for a time-related case they may be non-zero sources at some instant.

Furthermore, you may not connect two inductors in parallel. For the same, strict reason, each inductor is considered to be a voltage source. The parallel connection of two or more inductors forms a "voltage loop." You may avoid this restriction by including a series resistor with each inductor to "break" the loop. This may be a resistor of negligible value, say, 0.001 ohm, or one that accounts for the winding resistance (the DC resistance of the coiled wire), in which case it will have the same resistance value as the winding.

EXERCISE 2.8-1

Insert an inductor into one of the "legs" of the resistor bridge circuit (see earlier exercises). Run PSpice and note if the bias levels change. Try different "legs" of the bridge and check that the results were as you expected.

Exercise 2.8-2

By inserting two inductors in parallel, create a voltage loop in one of the "legs" of the resistor bridge circuit (see earlier exercises). Run PSpice and find the error message resulting from this situation.

CHAPTER 3

DC SENSITIVITY

Simulators are generally used either to verify a design or to refine (improve) a design. Verifying is simply checking that the design "meets spec." However, refining the design may make it more robust, attain a "tighter spec," or even make it less expensive to produce. DC sensitivity calculations help guide the user to those components that affect a circuit's DC bias-point the most. This, then, will focus efforts on reducing the sensitivity of the circuit to component variations and/or drift, or it may provide evidence that a design is too conservative and that less expensive components, with more variation and/or drift, may be used.

3.1 THE .SENS STATEMENT

The .SENS statement specifies which DC outputs you want to consider, since PSpice doesn't know how your circuit is being used. Then, once the DC bias-point for the circuit is calculated, PSpice calculates the sensitivity of each output, individually, to all the device values (as well as "model parameters," which we will cover in due course) in the circuit. The format for the statement is

.SENS *output_value* ...

The *output_value* is in the same format as for the .PRINT statement described in §4.2.

3.2 DC SENSITIVITY ANALYSIS

Having a .SENS statement in your circuit file triggers the DC sensitivity calculations after the DC bias-point calculations are completed. You do not need to specify any other output to get the results of the DC sensitivity analysis. We can now try working some examples to see what .SENS will do. Usually you will be using .SENS to analyze a more complicated, active device circuit, such as a transistor amplifier. However, for ease of understanding we can demonstrate the use of .SENS with a small demonstration circuit, and then work an example showing a practical application. (Remember, you **must** include the "." to start the statement.)
Consider the simple circuit

```
Resistor divider
Vin  1   0   1volt
R1   1   2   3ohm
R2   2   0   1ohm
.SENS V(2)
.END
```

which we will analyze. After running PSpice, the output file will look similar to

```
DC SENSITIVITIES OF OUTPUT V(2)
```

ELEMENT NAME	ELEMENT VALUE	ELEMENT SENSITIVITY (VOLTS/UNIT)	NORMALIZED SENSITIVITY (VOLTS/PERCENT)
R1	3.000E+00	−6.250E−02	−1.875E−03
R2	1.000E+00	1.875E−01	1.875E−03
Vin	1.000E+00	2.500E−01	2.500E−03

There will be a table like this for each *output_value* in the .SENS statement. Also, the sensitivities to selected currents would be labeled (AMPS/UNIT) and (AMPS/PERCENT).

What does this mean? First, let's look at the results for Vin. Our resistor divider has a voltage "gain" of 1/4; that is, the variation of the voltage at V(2) is one-quarter of the variation in Vin. This results in a sensitivity of a 0.25-volt change in V(2) for a 1.0-volt change in Vin, or 0.25 volts/unit (the "unit" for a voltage source being volts). How can we check this calculation? The .SENS statement also triggered the printing of bias-point calculation results. We can see that the ratio of V(2) to Vin is 1/4; the circuit has only linear elements, so doubling the value of Vin will double the value of V(2).

The last column shows sensitivity normalized to the component value, that is, as a percentage change. These values are then calculated by multiplying the former column (volts/unit) value by the element value, and then dividing by 100 to obtain a percentage value.

By looking at the values calculated for R2, which (again) are equal, since the resistor's value is 1.0 ohm, we can verify these by considering that

$$V(2) = \frac{R2}{R1 + R2}\, Vin \qquad\qquad (3\text{-}1)$$

This means that the normalized sensitivity of V(2) to R2 will be

$$\frac{\partial}{\partial R2}\, \frac{V(2)}{Vin} = \frac{(R1+R2) - R2}{(R1+R2)^2} = \frac{R1}{(R1+R2)^2} = \frac{3}{16} \qquad (3\text{-}2)$$

By analogy we can calculate the sensitivity of V(2) to R1 will be

$$\frac{\partial}{\partial R1}\, \frac{V(2)}{Vin} = \frac{(R1+R2) - R1}{(R1+R2)^2} = \frac{R2}{(R1+R2)^2} = \frac{1}{16} \qquad (3\text{-}3)$$

Notice that for R1 the normalized sensitivity is identical, in magnitude, to the normalized sensitivity for R2. It follows that a percentage change in either section of the resistor divider would produce the same-sized effect on the output. However, increasing R1 decreases the output voltage, so the sensitivity values are negative for this resistor.

EXERCISE 3.2-1

Run the sensitivity analysis just described. Now, change the value of R1 to 4 ohms and the value of VIN to 2 volts. Are the results what you expected?

3.3 CIRCUIT EXAMPLE: WORST-CASE DESIGN

One type of circuit where sensitivity to element values is of great importance is the digital-to-analog converter. These circuits generally use component ratios to generate a voltage that is a fraction of a reference voltage, where the fractional amount is set by a digital (binary-coded) input. Digital systems, such as computers, may generate analog signals by using these circuits. Consider the circuit in Figure 3-1, represented by the following circuit file:

```
* Sensitivity analysis of D-to-A converter
Vmsb 1  0  0volt; most-significant-bit input
Vlsb 2  0  0volt; least-significant-bit input
R1   1  3  20K
R2   2  4  20K
R3   3  4  10K
R4   4  0  20K
.sens v(3)
.end
```

The inputs to this circuit are set to either 1 volt, or zero, depending on the binary input we are simulating. In this type of converter, called an "R-2R ladder" because of the resistor ratios, the input voltages are both the binary input as well as the reference voltage. You may think of it as being the binary bit value, 1 or 0, multiplied by the reference voltage. The output voltage, at node 3, is a fraction of the reference voltage, controlled by the ratio of the current binary input value to the number of values representable. In this case, with 2 bits, we will be able to generate the following fractions: 0/4, 1/4, 2/4, and 3/4.

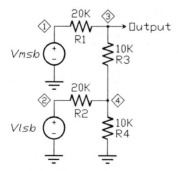

FIGURE 3-1 Schematic for worst-case example: D-to-A converter.

To generate all the input (binary code) cases we will need to make four runs. Each run will use a different combination of Vmsb and Vlsb, with values of either 0 volt or 1 volt, in the following combinations: 0-and-0, 0-and-1, 1-and-0, and finally 1-and-1. This provides the binary input for the decimal numbers 0, 1, 2, and 3. The output at V(3) will DC-bias to the voltage levels of 0, 0.25, 0.5, and 0.75, for these inputs, respectively. Looking at the sensitivity table for the inputs 0-and-0:

```
DC SENSITIVITIES OF OUTPUT V(3)
```

ELEMENT NAME	ELEMENT VALUE	ELEMENT SENSITIVITY (VOLTS/UNIT)	NORMALIZED SENSITIVITY (VOLTS/PERCENT)
R1	2.000E+04	0.000E+00	0.000E+00
R2	2.000E+04	0.000E+00	0.000E+00
R3	1.000E+04	0.000E+00	0.000E+00
R4	2.000E+04	0.000E+00	0.000E+00
Vmsb	0.000E+00	5.000E-01	0.000E+00
Vlsb	0.000E+00	2.500E-01	0.000E+00

These results are rather boring: since all the node voltages are zero it is difficult to affect the output voltage by changing a resistor value. Moving on, we look at the results for the combination 0-and-1:

```
DC SENSITIVITIES OF OUTPUT V(3)
```

ELEMENT NAME	ELEMENT VALUE	ELEMENT SENSITIVITY (VOLTS/UNIT)	NORMALIZED SENSITIVITY (VOLTS/PERCENT)
R1	2.000E+04	6.250E-06	1.250E-03
R2	2.000E+04	-7.813E-06	-1.563E-03
R3	1.000E+04	-6.250E-06	-6.250E-04
R4	2.000E+04	4.688E-06	9.375E-04
Vmsb	0.000E+00	5.000E-01	0.000E+00
Vlsb	1.000E+00	2.500E-01	2.500E-03

Then, we look at the table for the combination 1-and-0:

```
DC SENSITIVITIES OF OUTPUT V(3)
```

ELEMENT NAME	ELEMENT VALUE	ELEMENT SENSITIVITY (VOLTS/UNIT)	NORMALIZED SENSITIVITY (VOLTS/PERCENT)
R1	2.000E+04	-1.250E-05	-2.500E-03
R2	2.000E+04	3.125E-06	6.250E-04
R3	1.000E+04	1.250E-05	1.250E-03
R4	2.000E+04	3.125E-06	6.250E-04
Vmsb	1.000E+00	5.000E-01	5.000E-03
Vlsb	0.000E+00	2.500E-01	0.000E+00

Finally, we look at the table for the combination 1-and-1:

```
DC SENSITIVITIES OF OUTPUT V(3)
```

ELEMENT NAME	ELEMENT VALUE	ELEMENT SENSITIVITY (VOLTS/UNIT)	NORMALIZED SENSITIVITY (VOLTS/PERCENT)
R1	2.000E+04	-6.250E-06	-1.250E-03
R2	2.000E+04	-4.688E-06	-9.375E-04
R3	1.000E+04	6.250E-06	6.250E-04
R4	2.000E+04	7.813E-06	1.563E-03
Vmsb	1.000E+00	5.000E-01	5.000E-03
Vlsb	1.000E+00	2.500E-01	2.500E-03

Now we scan the tables for the resistors only, since we want to check the design against component variation. To calculate the normalized, worst-case deviation, we add the absolute value of the normalized deviations, for the resistors only, for each table. The absolute value is used because we assume that the resistors will deviate in the direction that changes the output voltage the most, *à la* "Murphy's Law." The table with the worst deviation is the third table, for input combination 1-and-0, repeated here showing the relevant rows and columns:

ELEMENT NAME	NORMALIZED SENSITIVITY (VOLTS/PERCENT)
R1	-2.500E-03
R2	6.250E-04
R3	1.250E-03
R4	6.250E-04

with a **maximum deviation** of 5mV for each percentage of allowable resistor value deviation. This suggests that for this digital-to-analog converter, a specification requiring 10mV maximum output deviation would allow the use of 2% (tolerance) resistors.

EXERCISE 3.3-1

Using the digital-to-analog converter example, now assume that the reference voltage (used at the inputs) has a 1% deviation. For each table, calculate the deviation due to the reference voltage, and subtract that result from a system specification of 10mV maximum deviation at the output. Using the remaining "allowable" deviation, what tolerance of resistors must be used to meet the system specification?

EXERCISE 3.3-2

Using the digital-to-analog converter example, assume that the resistor values track perfectly. What input combination gives the worst output deviation? (Hint: this is a trick question.)

CHAPTER 4

DC SWEEP

The simulations we have looked at so far calculated only quiescent, or DC, operation where the voltage or current sources maintained a fixed value. In this chapter we will look at circuits where the sources vary, though the analysis will still calculate quiescent (DC) operation. Using this type of analysis allows you to look at the results from many .OP analyses in a single simulation run. That is why it is called a "sweep."

Later in the chapter we will look at "controlled" sources. These allow you to build function blocks to transform signals.

4.1 SWEEPING A SOURCE: THE .DC STATEMENT

The DC sweep analysis is controlled with a .DC statement. When you "sweep" a source the simulator

- starts with one value for a source (voltage or current), then

- calculates the DC bias-point (exactly as it does for the .OP analysis), then

- increments the value and calculates another DC bias-point.

This increment-then-analyze procedure continues until the last source value has been analyzed. You get to select the starting value, increment, and final value for the sweep. The result is the same as doing many .OP analyses, but is faster if you want to check the range of source values due to

- new types of output available for this analysis, and

- the way the calculations are done.

The calculations for the DC sweep analysis are faster than the set of equivalent .OP analyses if only for the reason that PSpice does not have to reread the circuit file each time and then do the calculations. Beyond that, having arrived at the solution to the circuit for the initial source value, the solution for the next source value is assumed to be relatively close to the first solution. The first solution provides an estimate for the second solution. Then, having found the solution to the circuit for the first and second values, the solution for the third value is "guessed" by linear extrapolation from the first two solutions. This provides the estimate for the third solution. From then on, PSpice extrapolates from the previous two solutions for the next estimate solution (PSpice does not make use of more than the previous two solutions, because it has been found that, generally, the time required to calculate a higher-order extrapolation is not recouped by shorter solution times).

The .DC statement specifies the values used during the DC sweep. The statement says which source value is to be swept, the starting value, the amount to increment the value for each step of the sweep, and at what value to quit the sweep. In the syntax shown:

> .DC *source_name start_value stop_value increment_value*

the *source_name* is an independent source (voltage or current) in your circuit file. The .DC statement does not define the source, nor how it is connected to the circuit. The .DC statement says only what values that source will have during the DC sweep analysis. You need to make sure that you have specified the source in your circuit file, or PSpice will not be able to do the DC sweep analysis.

When adding a .DC statement to your circuit file, you do not need to change any of the other lines describing your circuit. Just add the .DC statement (remember, you **must** include the "." to start the statement), as the sweep of values specified will override the fixed value indicated by the independent source statement (V source or I source) during the DC sweep analysis (only).

4.2 SIMULATION OUTPUT: TABLES AND PLOTS

Being able to sweep through many values and calculate many results means you will want to get output that is different from the .OP analysis. Actually, it is the same output but the format has been changed so that it is more convenient to use. There are two features you would probably want to have in this new form of output; you should be able to:

- organize the calculated results so they could be referred to the value of the sweep source (like a table), and

- select which results are printed (to minimize the amount of output).

The .PRINT statement triggers the operation of selecting and tabulating results for the DC sweep analysis and other analyses to be explored later (and is used the same way for these other analyses, so learning it now will help later).

The .PRINT statement simply specifies which analysis the statement applies to, since it is used for many types of analyses, and which results to print. In the syntax shown, for printing results for the DC sweep:

> .PRINT DC *output_value...*

you can have many entries, or output values, in the table of results. Each *output_value* will get a column in the table, and each row of the column will be the calculated result of the output value for each step in the DC sweep. The columns are in the same order as specified in the .PRINT statement. (Remember, you **must** include the "." to start the statement.) Usually you will want to print the sweep value in the first column to simplify finding results in the table, so PSpice does this for you; the first column, which comes before the columns you specify, always contains the value of the sweep variable.

The output values you can print are basically node voltages and device currents (which also means source currents, as a source is also a device). Node voltages can be printed relative to ground (node "0") or relative to another node (that is, the value printed is the difference of the voltages at two nodes). The syntax is shown for the DC sweep:

.PRINT DC V(7)   to print the voltage at node 7

.PRINT DC V([out])   to print the voltage at node out (a "named" node)

.PRINT DC V(6,3)   to print differential voltage from node 6 to node 3

.PRINT DC V([a],4)   to print differential voltage from node a to node 4

.PRINT DC V(R1)   to print the voltage across R1 (any two-terminal device)

Device (and source) currents may be printed using the syntax

.PRINT DC I(R4)   to print the current through R4 (any two-terminal device)

You can, as mentioned earlier, print several values in one table, and mix voltages and currents; for example:

.PRINT DC V(3) V(4) I(R2)

EXERCISE 4.2-1

> Add a .DC statement to the bridge circuit of Figure 2-1 and sweep the supply voltage from 1 to 2 volts in 0.1-volt increments. Also add a .PRINT statement to print values for the supply (sweep) voltage and the current through two of the resistors. Does the table produced by PSpice indicate that this circuit is linear?

The calculated value of current through a device, such as a resistor, means positive current, which will be flowing from the more positive voltage level to the more negative voltage level. The value printed may have the opposite sign (that is, a negative value instead of positive) from the one you were expecting. This depends on the order of the nodes when you specified, say, a resistor, in your circuit file; the syntax "R4 3 5 150" means a 150-ohm resistor between nodes 3 and 5. If PSpice finds that node 3 is more positive than node 5, PSpice will calculate a positive (value) current through R4. You can think of current through the device as positive current flowing into the first node in the line specifying the device.

Exercise 4.2-2

> Using the previous example's circuit file, swap the nodes of one of the resistors specified in the `.PRINT` statement. How did the output change? Does this mean the circuit works any differently from before?

After a few simulations using the DC sweep analysis, you may notice that looking through the printed table is getting tedious. Wouldn't it be nice to have the computer graph the results? Well, PSpice will print graphs if you specify a `.PLOT` statement. Then you can use either `.PRINT` or `.PLOT` to look at the results, or both.

The `.PLOT` statement is nearly the same as the `.PRINT` statement. (Remember, you **must** include the "." to start the statement.) You specify the type of analysis the plot is for, and which results you want plotted. The output values have the same form as for the `.PRINT` statement. If you include two `.PLOT` statements in your circuit file you should get two plots. In the syntax for the DC sweep

`.PLOT` DC *output_value...*   [[*min_range , max_range*]]

notice that `.PLOT` will let you set the range of the output axis. If you do not specify the range, PSpice automatically calculates a range that includes all the output values.

Exercise 4.2-3

> Redo the previous exercise showing the use of the `.PRINT` statement with DC sweeps, but use a `.PLOT` statement instead.

4.3 Linear Controlled Sources

The controlled sources are one of the most useful, and overlooked, features of PSpice (and this is true of many other non-SPICE simulators). Controlled sources measure voltage or current and use the measured value to control their output, which can be either a voltage or current. The transformation allowed between input and output is a multidimensional polynomial. Both the number of dimensions (that is, number of measured inputs) and degree of each polynomial are set by you. But first, we will consider the linear case for these controlled sources before trying anything more difficult.

Allowing two types of input (voltage and current) and output yields four combinations of input/output:

- the voltage-controlled voltage source (VCVS)
- the current-controlled current source (CCCS)
- the voltage-controlled current source (VCCS)
- the current-controlled voltage source (CCVS)

These four sources are devices, just like resistors, and are given PSpice device types of E, F, G, and H, respectively. Mathematically, you may think of these devices as functions:

- VCVS is the function $v_{out} = E(v_1, v_2, v_3, \cdots)$
- CCCS is the function $i_{out} = F(i_1, i_2, i_3, \cdots)$
- VCCS is the function $i_{out} = G(v_1, v_2, v_3, \cdots)$
- CCVS is the function $v_{out} = H(i_1, i_2, i_3, \cdots)$

We specify these sources in a way that is similar to the passive devices we have been using: name, connecting nodes, and (instead of value) the transforming polynomial. As you might have guessed:

Exxx is a voltage-controlled voltage source

Fxxx is a current-controlled current source

Gxxx is a voltage-controlled current source

Hxxx is a current-controlled voltage source

In the most simple form, an example of a voltage-controlled voltage source

```
E2  5  7  3  4  10
```

is a voltage source, with output nodes of 5 and 7 (remember, the positive current is flowing out of the connection to node 5), and where the output voltage is controlled by the voltage present at nodes 3 and 4, with a simple multiplying gain of a factor of 10. This is the same as the equation $v_5 - v_7 = 10 \cdot (v_3 - v_4)$.

You may also want to include, for clarity, some superfluous (to PSpice) commas and parentheses to identify the input nodes. For example, the form

```
E2  5  7  (3,4)  10
```

makes the statement more device-like (a name, followed by two nodes, and then the value specification).

The same form may be used for the voltage-controlled current source, except that the multiplying value is converting voltage to current so it has the dimension of amps/volts. Instead of "gain" we have "conductance," but as there is a transfer from one set of nodes to another set, it is called "transconductance."

EXERCISE 4.3-1

Write down a VCCS statement for the function
$$i_3 = (v_5 - v_7) \div 20 \, \text{ohms}$$
What are the units for the transconductance value?

When the measured input is current, the syntax is different. PSpice needs to be told **which** current, that is, the current through which device. To simplify matters, PSpice measures currents through voltage sources (the fixed value V devices, not the variable

E or H devices described here). Instead of controlling nodes, the syntax includes the name of the V device that has the controlling current; for example:

```
F4 3 5 V2 5
```

is a current source whose output current is five times the current flowing through V2. Again, this is a simple amplification gain in current. If the device was an H device (CCVS) instead, there is a transformation from current to voltage. Then the units are volts/amps or "resistance," but again because of the transfer to another set of nodes, it is called "transresistance."

EXERCISE 4.3-2

Write down a CCVS statement for the function
$$v_3 - v_5 = \mathrm{I(V5)} \times 20 \, \text{ohms}$$
What are the units for the transresistance value?

You probably didn't realize that you have already used one of the controlled sources in the previous example circuits. To PSpice the resistor is simply a voltage-controlled current source, with the same input and output terminals! To the rest of a circuit

```
R5 3 2 120
```

is the same as

```
G5 3 2 (3,2) 120
```

The only difference is that when PSpice checks your circuit, the G device is a current source (with infinite impedance) and does not qualify as a DC path to ground.

EXERCISE 4.3-3

Review some of the previous exercises and replace resistors with G devices. Use .PRINT and/or .PLOT to verify that the operation of these circuits has not changed.

4.4 POLYNOMIAL CONTROLLED SOURCES

What is different about a polynomial controlled source (some call these "nonlinear" sources) versus the linear case is how the polynomial is described to PSpice. First, the dimension of the polynomial is specified. Then, the inputs to be measured must be described. Finally, the coefficients of the polynomial are specified. Single-dimension polynomial functions, which are basically the additive combination of many linear functions, are easily described using the syntax

```
{E|F|G|H}xxx node node POLY(1) inputs coefficient_list
```

As you may have guessed, polynomials in two dimensions (that is, having two controlling inputs) will use POLY(2) instead.

The coefficients in the *coefficient_list* are listed in order of ascending powers: that is, the list of coefficients a, b, c,... come from the formula $a + b \cdot x + c \cdot x^2 + \cdots$, where every coefficient up to the last non-zero coefficient must be specified. For example, if you wanted the formula

$$1 + 2x^3$$

you would specify the *coefficient_list*

    1   0   0   2

This way PSpice knows you are specifying a third-degree polynomial and that the coefficients of the higher degrees are all zero.

The controlling inputs come as pairs of nodes (for voltage inputs) or V device names (for current inputs). There must be as many pairs, or names, as there are dimensions to the polynomial. For now, we will focus on single-dimension polynomials.

EXERCISE 4.4-1

Create a circuit file that performs a DC sweep from -2 to $+2$ volts in 0.1-volt increments. Add to the file a VCVS that implements the function $x^3 - x$. Use .PRINT and .PLOT to check the output of the controlled source. Try this again with a VCCS.

When you want to have an output that is the sum of other input functions, for example:

$$\text{output} = f_1(x_1) + f_2(x_2) + \cdots$$

you do not need to use a higher dimension POLY(*ndim*); simply add the outputs. If they are voltage outputs, put the controlled sources in series. The voltages combine so that the voltage across the series is the sum of the individual voltages. The same is true for current outputs, except that you will want to have them in parallel with each other. This makes it easy to check the correctness of each function by itself before combining it with the other functions.

EXERCISE 4.4-2

Redo the previous exercise's function $x^3 - x$ as the sum of two functions: x^3 and $-x$. Run the DC sweep and check each function separately. Are there other ways to implement the $-x$ function?

4.5 Graphics Output†

For those of you used to the world of computer graphics, the output we have seen from the simulator seems primitive. What is missing is the ability to look at the response of your circuit on the computer display. After all, even some pocket calculators will plot graphic functions. For PSpice, we have such a facility called Probe™. To use Probe you must first tell PSpice to create a data file for Probe, which is done by using the `.PROBE` statement.

The `.PROBE` statement is very similar to the `.PRINT` and `.PLOT` statements. (Remember, you **must** include the "." to start the statement.) With `.PROBE` you may select node voltages and device currents to be output from the simulation. However, this is generally not the best use of the `.PROBE` statement; if you just put `.PROBE` in your circuit file, without specifying any particular outputs, PSpice will save **all** the node voltages and device (branch) currents. Then, later when you are using Probe to look at the results of the simulation, everything has been saved for your inspection. You select the waveforms you want to view. If you are curious about the operation of some section of your circuit, you may view it without rerunning the simulator (which is what you would need to do if the only output you had was `.PRINT` or `.PLOT`).

In the syntax shown, for saving results of any of the analyses

> `.PROBE` [[*output_value...*]]

remember that if you want to save all the voltages and currents from the simulation that you do not specify any output values. This "default" mode tells PSpice to save everything. If you do specify any output values, only those that are specified will be saved. You would normally do this only to save room in the data file, which PSpice is making as input for Probe.

The file created by PSpice for use by Probe is called by the same file name as the circuit file, but with the file extension .DAT (for example, the file MY.CIR would create MY.DAT). This Probe *data file* has a structure that tells Probe what analyses, voltages, currents, and independent variables (such as time, frequency, etc.) are available. Of course, the output data are in the file, too. For most of the circuits you will simulate, even when saving all the results, the Probe data files are quite modest in size by today's standards. Many complete circuit files, with the Probe data file output from PSpice, may be saved on a PC diskette. When you are saving a circuit, having a Probe data file for that circuit with all the output variables saved is valuable if you ever want to review the simulation results.

However, since PSpice saves outputs at each step during the simulation, long simulations of large circuits generate considerable output. Since you are usually interested in only a few voltages or currents, you can specify that PSpice saves only

---

† This section introducing graphics is placed here because it is helpful for you to be using waveform graphics for the rest of the book. Also, this is the first opportunity we have to look at any interesting waveforms.

those items to reduce the amount of data in the Probe data file. For large circuits, specifying a few output variables also speeds the simulation as PSpice skips the calculations and file-buffering operations necessary to save all the other output variables.

Probe is simple to use and is menu driven, so you don't need to remember any commands or statements (as with PSpice). Most of your difficulty will probably come from getting Probe set up and started the first time. Then it will be easy and you can focus on your simulations.

EXERCISE 4.5-1

Put a .PROBE statement in the circuit file for the sweep of the function x^3-x. Use Probe to examine the output you had plotted using the .PLOT statement.

EXERCISE 4.5-2

What other ways could you generate or display the curve of x^3-x?

4.6 MULTIPLE-INPUT CONTROLLED SOURCES

When the controlled source you want is a function of several inputs, describing the coefficients can become complicated. In general there are many possibilities, and the general form must allow for all of them. PSpice is told only the number of dimensions (that is, the number of "measuring" inputs) for the polynomial, so the list of coefficients follows a rule to describe the function you want. The general form description is complex to follow, but we can look at a three-input case to get the general idea. Assuming our inputs are v_1, v_2, v_3, and a list of coefficients called k_0, k_1, k_2,... the polynomial form for three inputs is

the constant term	k_0
plus, the linear terms	$k_1 \cdot v_1 + k_2 \cdot v_2 + k_3 \cdot v_3$
plus, cross terms	$(k_4 \cdot v_1 + k_5 \cdot v_2 + k_6 \cdot v_3) \cdot v_1 +$ $(k_7 \cdot v_2 + k_8 \cdot v_3) \cdot v_2 +$ $(k_9 \cdot v_3) \cdot v_3$
plus, more cross terms	$\begin{pmatrix} (k_{10} \cdot v_1 + k_{11} \cdot v_2 + k_{12} \cdot v_3) \cdot v_1 + \\ (k_{13} \cdot v_2 + k_{14} \cdot v_3) \cdot v_2 + \\ (k_{15} \cdot v_3) \cdot v_3 \end{pmatrix} \cdot v_1 +$ $\begin{pmatrix} (k_{16} \cdot v_2 + k_{17} \cdot v_3) \cdot v_2 + \\ (k_{18} \cdot v_3) \cdot v_3 \end{pmatrix} \cdot v_2 +$ $\Big((k_{19} \cdot v_3) \cdot v_3\Big) \cdot v_3$

and so on for every combination of inputs. Obviously it is easy to make errors if we have many inputs! Fortunately this rarely happens.

A simple use of the general case is to sum several input voltages. A four-input voltage summer would have the form

```
Eout 7 0 poly(4) (1,0) (2,0) (3,0) (4,0) 0 1 1 1 1
```

Again, notice the parentheses around the voltage node pairs, which are also comma separated. You may do this in PSpice, and most SPICEs, as the commas and parentheses are treated like spaces. This improves the "readability" of the polynomial form.

EXERCISE 4.6-1

Write down a voltage summer with the following weighted inputs:
$$v_1 + 3v_2 + 2v_3 + 0.5v_4$$

Another common case is a two-input controlled source. You may have thought about how to multiply two voltages. Using the syntax described above, a voltage multiplier would be

```
E2 3 4 poly(2) (7,8) (5,6) 0 0 0 0 1
```

where the output voltage (across nodes 3 and 4) is a function of the input voltages (across nodes 7 and 8, and nodes 5 and 6). The offset coefficient (that is, the coefficient for the zeroeth degree) is zero, as well as the coefficients for linear voltage terms. That's three zeros, so far. The fourth zero is for the quadratic voltage term of the first input voltage. Then we arrive finally at the coefficient for the multiplication of the two input voltages, which is set to unity.

Arriving at a cubic function of two inputs is too painful using the general syntax. It is easier for you to decompose the function into two stages (if this is possible); for example:

$$x \cdot y^3$$

becomes

$$(x) \cdot (y^3)$$

This way you create the individual functions (and test them, if you are uncertain about their operation), and then combine the intermediate outputs with the multiplier described previously.

4.7 FUNCTION MODULES

By using the controlled sources you can create a variety of modular function blocks. With a set of these in your "simulator toolkit," you can quickly check circuit ideas. And, even though we are investigating these using DC sweep analysis, these blocks also work for all the other types of analyses.

We have already covered the voltage-multiplier function using a controlled-source statement

$$\text{Exxx } +node\ -node\ \text{POLY(2) } +Anode\ -Anode\ +Bnode\ -Bnode\ 0\ 0\ 0\ 0\ 1$$

which will multiply the voltages across the node pairs *A* and *B*. If you were to connect the same nodes to both the *A* and *B* inputs of the voltage multiplier, the output would be the square of the input voltage. This is a trivial extension of the multiplier; however, if we include the multiplier in a feedback loop, we can develop new uses.

EXERCISE 4.7-1

Create a current multiplier that multiplies the current flowing through two independent voltage sources. Test it to be sure the direction of current output is what you expected.

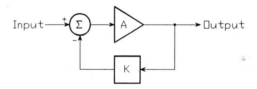

FIGURE 4-1 Schematic of generalized feedback circuit.

Feedback theory tells us that a circuit, as shown in Figure 4.1, with three major sections:

- a forward path, including a perfect (one hopes) amplifier,
- a feedback path, which has the interesting circuitry, and
- a difference block, which creates the "error" signal

will develop the transfer function

$$\text{output} = \frac{A}{A{\cdot}K + 1}\ \text{input} \tag{4-1}$$

so that the gain of the circuit is

$$\text{gain} = \frac{\text{output}}{\text{input}} = \frac{A}{A{\cdot}K + 1} = \frac{1}{K}\ \frac{A{\cdot}K}{A{\cdot}K + 1} \tag{4-2}$$

so that if the forward amplification gain, A, is large, the circuit gain becomes 1/K.

With the controlled-source statement we can easily create "perfect amplifiers" that are linear and have huge gain ratios. In fact, we can even integrate the amplifier with the difference function to create a "perfect error amplifier" by implementing the function

$$\begin{aligned} \text{output} &= \text{gain} \cdot (\text{input} - \text{feedback}) \\ &= \text{gain} \cdot \text{input} - \text{gain} \cdot \text{feedback} \end{aligned} \tag{4-3}$$

Now for the feedback section. If the entire circuit were to perform, say, the square-root function, then the function the feedback section has to perform is

$$\text{gain} = \frac{\text{output}}{\text{input}} = \frac{\sqrt{\text{input}}}{\text{input}} = \frac{1}{K} \tag{4-4}$$

This means the feedback section has to square the output voltage, a function we covered earlier. Now we can build a circuit that calculates, with a small degree of error, the square-root of the input voltage:

```
Square-root circuit
Vin  1 0 0
Rin  1 0 1E6
Efwd 2 0 poly(2) (1,0) (3,0) 0 1E6 -1E6 ; error amplifier
Rfwd 2 0 1E6
Erev 3 0 poly(2) (2,0) (2,0) 0 0 0 0 1  ; feedback section
Rrev 3 0 1E6
.DC Vin 0 10 .1
.PROBE
.END
```

Notice that resistors were placed across all the source outputs. Without them the source outputs would be dangling, for the inputs to the controlled sources are considered by PSpice to have infinite impedance.

EXERCISE 4.7-2

Build and run the square-root circuit. Check the output values. How could the feedback section be simplified to have only one input? Try a different DC sweep, starting at +5 volts and sweep to −5 volts. Why doesn't the circuit work for negative input voltages?

EXERCISE 4.7-3

Build and test a cube-root circuit. Does it work for negative input voltages?

Now suppose we wanted a circuit that produces an output equal to the ratio of the two input voltages. Then the function the feedback section has to perform is

$$\text{gain} = \frac{\text{output}}{\text{input}_1} = \frac{\left(\dfrac{\text{input}_1}{\text{input}_2} \right)}{\text{input}_1} = \frac{1}{\text{input}_2} = \frac{1}{K} \tag{4-5}$$

This means the feedback section has to multiply the output voltage by the denominator point. Now we can build a circuit that calculates, with a small degree of error, the ratio of two input voltages:

```
Divider circuit
Vtop 1 0 1 ; top of fraction
Rtop 1 0 1E6
Vbot 2 0 1 ; bottom of fraction
Rbot 2 0 1E6
Efwd 3 0 poly(2) (1,0) (4,0) 0 1E6 -1E6 ; error amplifier
Rfwd 3 0 1E6
Erev 4 0 poly(2) (3,0) (2,0) 0 0 0 0 1 ; feedback section
Rrev 4 0 1E6
.DC Vbot .1 1 .05
.PROBE
.END
```

EXERCISE 4.7-4

Build and test the divider circuit. Try extending the function to be the ratio of two independent, second-order polynomials.

4.8 SUBCIRCUITS

Now that we have built some useful function blocks, we still have the problem that, for each use of a block, we need to re-key the section of circuit into our circuit file. Furthermore, if we wanted multiple blocks of the same type, unique nodes and component names would have to be used. This would quickly become tedious and inflexible. To help out, PSpice has a macro facility called a "subcircuit" that captures a circuit function as a "sub-network" of connected components. Terminals are assigned for the subcircuit to connect it into your circuit. The subcircuit definition has the form

```
.SUBCKT name node [[node...]]
    components defining subcircuit
.ENDS
```

where the list of nodes identifies which nodes of the subcircuit are terminals that may be attached to the external circuit (an enhanced form of this statement, which accepts parameter values, is described in §10.7). Node numbers within the subcircuit are separate from, and are not to be confused with, any nodes that might be in the external circuit (except for the "0" or "ground" node, which is a global node reference). For example, we could encapsulate the square-root function from before by using the following subcircuit definition:

```
.SUBCKT SqRoot 1 2 3 4
  Efwd 3 4 poly(2) (1,2) (5,0) 0 1E6 -1E6; error amplifier
  Erev 5 0 poly(1) (3,4) 0 0 1; simplified feedback section
  Rrev 5 0 1E6
.ENDS
```

Notice that the input nodes (1 and 2) and the output nodes (3 and 4) are "floating," so that the function block is now similar to the controlled-source statements and may be inserted in any circuit.

To use the subcircuit in a circuit file PSpice uses the convention that these "new" devices are treated as a new device type, with names starting with an X, as follows:

   Xxxx node [[node...]] name

Each of the terminal nodes in the subcircuit definition must be used when the subcircuit is used, and the order of the nodes is the same as the order in the definition of the subcircuit. For example, we will rewrite the test circuit for the square-root function from before:

```
Square-root test circuit
Vin 1 0 0
Rin 1 0 1E6
Xblock 1 0 2 0 SqRoot ; here is where we use the subcircuit
Rout 2 0 1E6

*Square root definition
.subckt SqRoot 1 2 3 4
  Efwd 3 4 poly(2) (1,2) (5,0) 0 1E6 -1E6 ; error amplifier
  Erev 5 0 poly(1) (3,4) 0 0 1; simplified feedback section
  Rrev 5 0 1E6
.ends

.DC Vin 0 10 .1
.PROBE
.END
```

The subcircuit definition may occur anywhere between the title line and the .END statement. With PSpice you may collect all of these function blocks into a "toolkit" library file that will be searched by the command

   .LIB   *"file_name"*

to bring in only the subcircuit definitions needed by a particular circuit file. (Remember, you **must** include the "." to start the .LIB statement.) You can use these library files to save and reuse function blocks that are useful to your work.

The "transfer function" analysis is another DC-bias analysis and is used to calculate some external, or "black box," characteristics of your circuit. For this analysis, the outputs represent values for

- "small-signal" DC gain (input-to-output transfer ratio),

- DC input resistance, and

- DC output resistance.

We will examine what these values tell you about your circuit.

5.1 SMALL-SIGNAL DC ANALYSIS

The transfer function of your circuit involves "small-signal" DC analysis. "Large" signals are the normal excursions your circuit might encounter; for example, the output fluctuations of a stereo amplifier are generally large. "Small" signals are minuscule; for example, the signal amplified by a radio receiver is quite small. However, "small-signal" analysis deals with circuit operation in the limit of signals approaching zero strength. When your circuit has linear operation for the signals it will normally encounter, small-signal calculations may be applied to an operation that is merely "small-ish." Large signals need to be treated in a different fashion, which we will get to in due course.

Even though small-signal analysis is for small (nearly zero level) signals, this does not mean that your circuit is "turned off" and all the nodes are at zero volts. It just means that the stimulus to the circuit is small, but the power supply, for example, has its normal value. When your circuit is energized by DC sources, but there is no other external stimulus, this is the same as the bias-point situation we simulated before. Small-signal DC analysis, then, performs a calculation of the effects of minuscule input stimulation to your circuit. It answers, for example, the question "If the input node were to deviate slightly from its bias-point value, what would the output do?"

5.2 Circuit Gain

The most common question about the small-signal operation of a circuit is, "What is the gain?" This is probably because the most common electronic circuit is an amplifier (even digital circuits amplify), and its gain is a fundamental specification. Gain is the ratio of output signal deviation to input signal deviation. We use the word "deviation" because we need to differentiate between the quiescent, or steady-state, level of the input and the small excursions that represent the "real" signal. Mathematically, small-signal DC gain is the derivative of output with respect to input, at the DC bias-point and at **zero** frequency (you might think of this as the small-signal AC gain in the limit as frequency goes to zero). For example:

$$\frac{\partial v_{out}}{\partial v_{in}} \tag{5-1}$$

is an expression for *voltage gain*. There are other types of "gain" you may be interested in. As you might have guessed

$$\frac{\partial i_{out}}{\partial i_{in}} \tag{5-2}$$

is an expression for *current gain*. However, you may also want to evaluate

$$\frac{\partial i_{load}}{\partial v_{in}} \tag{5-3}$$

which is an expression for *transconductance*, or

$$\frac{\partial v_{out}}{\partial i_{in}} \tag{5-4}$$

which is an expression for *transresistance*.

5.3 Input and Output Resistance

Resistance, as you will recall from Ohm's law, is the ratio of voltage across the resistor to current flowing through the resistor: $V = I \cdot R$ or $R = V/I$. Input or output resistance of a circuit is much the same, although we are now referring to the "dynamic," or small-signal, resistance at the input or output. Mathematically, small-signal DC resistance is the derivative of the input voltage with respect to the input current, at the DC bias-point (and at zero frequency). For example:

$$\frac{\partial v_{in}}{\partial i_{in}} \tag{5-5}$$

is an expression for *input resistance*.

5.4 THE .TF STATEMENT

The `.TF` statement specifies what you consider to be the "input" and "output" of your circuit (PSpice doesn't know how your circuit is being used). Once the DC bias-point for the circuit is calculated, PSpice calculates the following "black box," or "transfer" functions: gain, input resistance, and output resistance. The format for the statement is

> `.TF` *output_variable input_source_name*

where the *output_variable* is in the same format as for the `.PRINT` statement. The *input_source_name* must be an independent source (V or I device); this is because the input usually has some fixed input bias, even if it is zero, which you may want to set (also, remember that PSpice will not analyze circuits with "dangling" nodes).

5.5 TRANSFER FUNCTION ANALYSIS

Having a `.TF` statement in your circuit file triggers the transfer function calculations. These are done when the DC bias-point calculations are completed. Just as for the `.OP` statement, you do not need to specify any other output, such as `.PRINT`, to get the results of the transfer function analysis. We can now try working some examples to see what `.TF` will do. Usually you will be using `.TF` to analyze a more complicated, active device circuit, such as a transistor amplifier. However, for ease of understanding we can demonstrate the use of `.TF` with linear and nonlinear controlled sources as our active elements. (Remember, you **must** include the "." to start the statement.)

Consider the simple circuit

```
Resistor divider
Vin 1 0 1volt
R1   1 2 3ohm
R2   2 0 1ohm
.TF V(2) Vin
.END
```

which we will analyze. We run PSpice, and in the output we find (something similar to):

```
**** SMALL-SIGNAL CHARACTERISTICS

     V(2)/Vin =  2.500E-01
     INPUT RESISTANCE AT Vin =  4.000E+00
     OUTPUT RESISTANCE AT V(2) =  7.500E-01
```

What does this mean? Since we specified both input and output variables as a voltage at a node and a voltage source, respectively, the transfer function gain calculation is voltage gain. Our resistor divider has a voltage gain of 1/4, meaning that the variation of the voltage at $V(2)$ is one-quarter of the variation in VIN. The input resistance "seen" by (input) VIN is 4 ohms, as you might have expected. This means that a variation in the voltage (in volts) of VIN will be four times the measured current variation (in amps) of VIN as a result of the voltage variation. The output resistance "seen" by (output) node 2 is 3/4 ohm. This means that if you could manage to vary the voltage at node 2, the size of the variation (in volts) would be three-quarters the size of the current (in amps) required from the means by which you accomplished the variation. How can we check these calculations?

Checking the value for gain is fairly simple; the .TF statement also triggered the printing of the bias-point calculation results. We can see that the ratio of V(2) to VIN is 1/4; the circuit contains only linear elements, so doubling the value of VIN doubles the value of V(2).

EXERCISE 5.5-1

Run the transfer function analysis just described. Suppose you don't immediately see that doubling the value of VIN will double the value of V(2) for the DC bias-point... try it.

Checking the input resistance "seen" by VIN is elementary, too. The output from PSpice also included the current supplied by VIN. Since VIN is the only source of current to the circuit, we may divide VIN's voltage by its supply current to arrive at the resistance "seen" by VIN. You might be concerned that the literal answer to this formula is −4 (ohms); however, the −0.25 amp current in the output file indicates merely that the current is flowing out of VIN's first node. This is from the SPICE convention for current direction: positive currents flowing into a terminal have a positive value.

Checking output resistance is easy, too. Since it was stated that output resistance was the ratio of the output voltage change to the current of an external influence, let us try that. By adding the current source

```
IOUT 2 0 0.1amp
```

and re-simulating, we can "draw down" the output voltage. Now we see that the voltage at V(2) is 0.175 volt, instead of the 0.25 volt it was before. The output resistance is calculated as

$$\frac{0.25-0.175}{0.1} = 0.75 \tag{5-6}$$

From the perspective of node 2, R1 and R2 are in parallel because the voltage source VIN is ideal and has zero resistance. You could have calculated the output resistance as the parallel combination of R1 and R2:

$$\frac{1}{\dfrac{1}{R1}+\dfrac{1}{R2}} = \frac{1}{\dfrac{1}{3}+\dfrac{1}{1}} = \frac{3}{1+3} = \frac{3}{4} = 0.75 \tag{5-7}$$

EXERCISE 5.5-2

Try the method just described to check the output resistance. Now, instead of using a current source, use a voltage source to set the voltage of V(2) to 0.3 volt. How much current was supplied by the new voltage source? How can you use this information to calculate output resistance? Did you get the same value?

5.6 LINEAR EXAMPLE

What kind of transfer function information is calculated for a circuit with linear gain? Consider the circuit in Figure 5-1, represented by the following circuit file:

```
Simple gain-of-5 circuit
Vin   1 0 1volt
Rin   1 0 1ohm
Gout  0 2 (1,0) 5.0
Rout  2 0 1ohm
.TF V(2) Vin
.END
```

This circuit is an ideal "gain block" with input and output resistance (both are 1 ohm). Gout was connected so that positive current flows into node 2, making the gain of the circuit a positive value; when V(1) increases in value, V(2) also increases in value. The result of the simulation will be something similar to

```
**** SMALL-SIGNAL CHARACTERISTICS

    V(2)/Vin =   5.000E+00
    INPUT RESISTANCE AT Vin =   1.000E+00
    OUTPUT RESISTANCE AT V(2) =   1.000E+00
```

These are the results we expected.

EXERCISE 5.6-1

You may have noticed that, in the previous example, Vin was set at 1 volt. Does this matter? Try setting it to a different value, say 2 volts, and rerun the simulation. What changed? Why didn't the output of the transfer function analysis change?

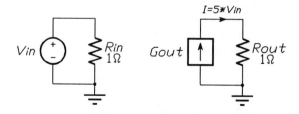

FIGURE 5-1 Schematic for gain-of-5 circuit example.

Exercise 5.6-2

Assuming you know about Thévenin equivalent transformations, change the example above to use an E device as the gain element. Check your work by running the simulator to see that the same values are calculated for the transfer function analysis. (Hint: your output resistance may not "dangle," so connect the output node to ground with a large resistance.)

5.7 Nonlinear Example

Most electronic circuitry is nonlinear; however, the transfer function analysis performs linear calculations on the circuit. That is, the calculations are done once the circuit has attained its bias-point and has been "linearized." This means that all of the elements in the circuit are expressed as their linear equivalents, which are valid only for that particular bias-point. A different bias-point would probably require a different "linearization." So far we have looked only at examples that were already linear, so the process of linearizing these elements did not change them. Consider the following example, which has a nonlinear element:

```
Nonlinear gain circuit
Vin  1 0 1volt
Rin  1 0 1ohm
Gout 0 2 poly(1) (1,0) 0 0 1
Rout 2 0 1ohm
.TF V(2) Vin
.END
```

This circuit is very much like the previous example, which was linear. However, the gain element is nonlinear, as its output is the square of the voltage at node 1. After running the simulator, the output will be something similar to

```
**** SMALL-SIGNAL CHARACTERISTICS

     V(2)/Vin =  2.000E+00
     INPUT RESISTANCE AT Vin =  1.000E+00
     OUTPUT RESISTANCE AT V(2) =  1.000E+00
```

The input is 1 volt, and the output is 1 volt (the input voltage squared), but why is the gain calculated as a value of 2.0? This is due to the linearization of the circuit at the bias-point. The linear slope value of the gain circuit is the mathematical derivative of the gain function. Remember, we defined small-signal voltage gain as $\partial v_{out}/\partial v_{in}$ and now you can see the result of this definition. How can we check this value for gain? Rerun the simulation with Vin set to 1.01 volts. This time the values printed for the bias-point will show V(2) as 1.0201 volts, so the gain of the circuit for small excursions in the input voltage is

$$\frac{\Delta V(2)}{\Delta Vin} = \frac{1.0201 - 1}{1.01 - 1} = 2.01 \tag{5-8}$$

You can see that, for small deviations in the input voltage, the output deviations are twice as large. In this case, PSpice shows enough precision in the printed values to see the effect of the square function; we can relate the output from the last two simulations to what we know about *Taylor series* from mathematics.

Taylor series involve the mathematical linearization of functions. The series allows you to calculate values for the function in the neighborhood of a point on the function (which we call the "bias-point" in electronics). If $f(x)$ has continuous derivatives in the region of a point $x = a$, then

$$f(x) = f(a) + \frac{(x-a)}{1!} f'(a) + \frac{(x-a)^2}{2!} f''(a) + \cdots \qquad (5\text{-}9)$$

From the first simulation of the "voltage squared" we have the result for $a = 1.0000$ and $f(a) = 1.0000$; these are the bias-point node voltages. Now, to predict the results of the second simulation, with $x = 1.01$, we know

the original function: $f(x) = x^2$

its first derivative: $f'(x) = 2 \cdot x$

its second derivative: $f''(x) = 2$

so the Taylor series calculation for $a = 1.00$ and $x = 1.01$ is

$$f(x) = 1.00 + \frac{(1.01 - 1.00)}{1!} \cdot 2 \cdot 1.00 + \frac{(1.01 - 1.00)^2}{2!} \cdot 2$$
$$= 1.00 + 0.02 + 0.0001 \qquad (5\text{-}10)$$
$$= 1.0201$$

which is the same result PSpice calculated for $V(2)$ in the second simulation.

EXERCISE 5.7-1

The circuit above is a "voltage squared" function. What transfer function results would you expect if Vin were -1.0 volts? What bias-point would you expect if Vin were -1.01 volts? Check this with the simulator and by applying the Taylor series formula.

EXERCISE 5.7-2

Try similar simulations for the function $f(x) = x^3$.

5.8 PLOTTING SMALL-SIGNAL GAIN

Probe graphics can be used to plot gain over a sweep of DC operating conditions. Running several simulations with .TF analysis is tedious. Fortunately this technique does not use the .TF at all, but is included here as an alternate way to get small-signal gain numbers from PSpice. Note that this is for gain only. While there is an equivalent technique for sweep results of input or output resistance, it is much more interesting to look at sweeps of frequency. We will look at this later in the book.

The definition of small-signal voltage gain is $\partial v_{out}/\partial v_{in}$. Any of the variables or formulas you enter into Probe may be differentiated by wrapping them in the d( ) function. Also, single voltage or current values may be differentiated by using a d prefix. For example, the derivative of V(2) is dV(2). Probe is able to calculate approximations for derivatives by using divided differences. This follows the notion that in the limit, as a approaches x

$$\lim_{a \to x} \frac{f(x-a)}{x-a} = f'(x) \tag{5-11}$$

If we are careful to make the differences small enough, the calculations should be useful. (We also should not make the differences too small, or Probe could lose accuracy in its calculations.)

Using the "voltage squared" circuit from earlier in this chapter, we can look at its small-signal gain over a range of bias conditions using the DC sweep and Probe. The circuit would be set up as shown:

```
Nonlinear gain circuit
Vin   1 0 1volt
Rin   1 0 1ohm
Gout  0 2 poly(1) (1,0) 0 0 1
Rout  2 0 1ohm
.DC Vin -2 2 .05
.PROBE
.END
```

After the simulation is finished, start Probe. (You may have this set up to happen automatically.) The divided differences mentioned above are calculated relative to the

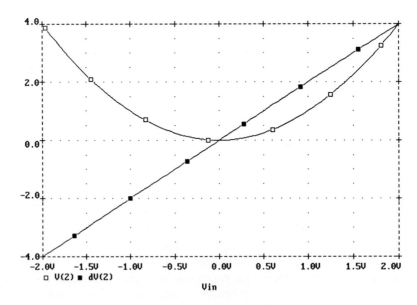

FIGURE 5-2 Plot of small-signal gain.

displayed X-axis. For example, if `Vin` is the X-axis, the Probe function `dV(2)` is shorthand for $\partial V(2)/\partial Vin$. By displaying the variables `V(2)` and `dV(2)`, we see the plot of small-signal gain shown in Figure 5-2, which indicates the output voltage and the small-signal gain (approximately) for the circuit.

EXERCISE 5.8-1

Try setting the step size of the DC sweep to a larger value, say, 0.2 volts, and see how the graph of `V(2)` and `d(V2)` changes.

EXERCISE 5.8-2

Try the previous example and exercise with `Gout` as a cubic function.

One of the more popular uses of circuit simulators is to verify the frequency response of signal filter and control circuits. The frequency response analysis calculates all of the AC node voltages and branch currents over a swept range of frequencies. The output of this analysis are the values for

- the amplitude of node voltages and device currents, and

- the relative phase angles of the node voltages and device currents.

By "frequency response" we mean "small-signal frequency response," where the analysis is done with the assumption that the input signals are small enough to minimize nonlinear effects. Novice users of PSpice often confuse frequency response with "transient response," which we will look at later. They relate frequency response to a lab-bench setup that sweeps the input of their circuit, with an oscillator, while they look at the output with an oscilloscope. While it is true the oscillator is providing a waveform with a frequency, often the signal is large enough to induce nonlinear behavior in the circuit. This type of experiment may be simulated also, but not by using frequency response analysis.

At the beginning of frequency response (AC) analysis, the DC bias-point for the simulated circuit is calculated. PSpice uses the same procedure as for the DC bias-point (.OP) analysis. Then the linear component equivalent of the circuit is "saved" (this is what is meant by saying the circuit was linearized) and used for the AC analysis. The laws of Ohm and Kirchhoff apply for AC analysis, too, but the impedances between nodes are said to be *complex* (having both *real* and *imaginary* components), which provides different results from DC analysis.

6.1 SPECIFYING INPUT SOURCES

From before, you recall that the independent voltage sources (V device) and current sources (I device) had the statement form

> *name node node value*

where *value* was the DC voltage or current level, depending on the device type. Actually, the *value* part of the statement could be stated in the form

> DC *value*

to remind you that the value is a DC bias amount; however, PSpice allows you to include only the numeric value, and it is implied that this is the DC value. Reiterating, the statement

```
VIN 1 0 0.5
```

is "shorthand" for

```
Vin 1 0 DC 0.5volt
```

and both represent the same voltage source.

We cover this shorthand because it may only be used for the DC value, but **not the AC value**. PSpice can allow a shorthand for only one of the specifications. A more complete representation of the input source statement is

name node node [[*DC-value*]] [[*AC-value*]]

where if you leave out *DC-value*, the DC value is set to zero. Likewise for *AC-value*... if you leave it off, the AC value is zero. You may even leave off both values, so that

```
VIN 1 0
```

is "shorthand" for

```
Vin 1 0 DC 0.0volts AC 0.0volts 0.0degrees
```

and both represent a "zero volt" source (which we have used before as a current monitor). You will include values for both in situations where you want to use an independent source that has both a DC and an AC value (the AC signal rides on a DC input level).

The *AC-value* portion of the statement has the form

AC *magnitude* [[*phase*]]

where you may leave off the *phase* value if the phase is zero. The *magnitude* value is straightforward and is simply the peak amplitude of the AC excitation. The *phase* value is the offset phase you want to have for this source, and the offset amount is relative to "zero phase." This may seem excessively general for most circuits where the output phase is already relative to the input phase, so there is usually no need to shift the input phase. However, PSpice can deal with multiple input sources of differing magnitudes and relative phases, so you may need to shift phase for a more complicated circuit.

Note that, since the circuit has already been linearized for this analysis, the excitation level you choose is arbitrary. The levels calculated for the rest of the circuit will change in proportion to the input. For this reason we normally set the input magnitude to unity so that all of the calculated levels represent "gain."

6.2 THE .AC STATEMENT

The .AC statement specifies the frequency values used during the frequency response analysis. The statement says only which frequencies are used and the **AC values for all independent sources will be set to these frequencies.** Usually there is only one source that has a non-zero AC value, and it becomes the input source of AC signal. However, PSpice is **not** limited to just one AC signal source.

The frequency sweep comes in three types: linear, octave, and decade. Their syntax forms are similar. The statement

.AC LIN *points start stop*

defines a linear frequency sweep, with *points* specifying the number of points in the sweep starting at the *start* frequency and finishing at the *stop* frequency. The statement

.AC OCT *points start stop*

defines a logarithmic frequency sweep, with *points* specifying the number of points per octave (an octave represents a twofold increase in frequency) in the sweep starting at the *start* frequency and finishing at the *stop* frequency. The statement

.AC DEC *points start stop*

defines a logarithmic frequency sweep, with *points* specifying the number of points per decade (a decade represents a tenfold increase in frequency) in the sweep starting at the *start* frequency and finishing at the *stop* frequency. (Remember, you **must** include the "." to start the statement.)

6.3 PRINT AND PLOT OUTPUT

Output from AC analysis may be generated by .PRINT or .PLOT statements, just as in DC analysis. In either case, the output is organized by the frequency at which the calculations were made. The statement forms are

.PRINT AC *output...*

and

.PLOT AC *output...*

Each *output* entry becomes a column in the table output by the .PRINT statement or a curve in the plot output by the .PLOT statement. The output values you can print/plot are node voltages and device currents (which also means source currents, as a source is also a device) with some special considerations for AC analysis. Node voltages and device currents may be specified as magnitude, phase, real, or imaginary, plus some functions, by adding a suffix to V or I:

(no suffix)	for magnitude
M	for magnitude
DB	for magnitude in decibels: $20 \times log$(value)
P	for phase
G	for group delay (not in SPICE)
R	for real part
I	for imaginary part

So, for example:

`.PRINT AC V(7)`	prints the voltage magnitude at node 7
`.PRINT AC V([out])`	prints the voltage magnitude at node `out` (a "named" node)
`.PRINT AC VP(7)`	prints the voltage phase at node 7
`.PRINT AC IR(R1)`	prints the real part of the current through `R1`

You may print several values in one table, and mix voltages and currents, as in this example:

```
.PRINT AC V(3) VP(4) II(R2)
```

Usually you will want to print the analysis frequency in the first column to simplify finding results in the table, so PSpice does this for you: the first column, which comes before the columns you specify, always contains the value of the analysis frequency.

6.4 GRAPHICS: MAGNITUDE AND PHASE PLOTS

Using Probe with AC analysis is identical to the way we used it with the DC sweep; just include a `.PROBE` statement in the circuit file. Let's try a small filter circuit to explore the kinds of result we can get from the simulator.

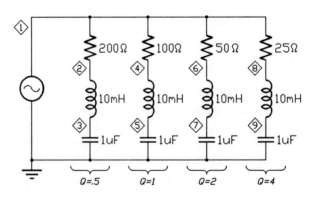

FIGURE 6-1 Schematic of low-pass filter example.

Figure 6-1 shows the schematic of a double-pole, low-pass LC-filter circuit. The only unusual feature about this circuit is that we have split the input resistance into several sections. This will allow us to investigate the response of the circuit with different values of input resistance, all in one simulation. The equivalent circuit file is

```
Four double-pole, low-pass, LC-filters
VIN 1 0 AC 1
* Q = .5
R1  1 2 200
L1  2 3 10mH
C1  3 0 1uF
* Q = 1
R2  1 4 100
L2  4 5 10mH
C2  5 0 1uF
* Q = 2
R3  1 6 50
L3  6 7 10mH
C3  7 0 1uF
* Q = 4
R4  1 8 25
L4  8 9 10mH
C4  9 0 1uF
.AC DEC 100 100hz 10Khz
.PROBE
.END
```

Note that the input level selected for the AC source is 1 volt. Since frequency response analysis is a small-signal analysis, this simplifies looking at the ratio of output response to input response. If the input equals 1, there is no need to literally calculate the ratios, since the output value is the ratio.

The circuit values were selected to have a resonant frequency in the audio range and a different quality factor "Q" for each circuit section. This provides an interesting selection of circuit responses, from which you might derive a qualitative "feel" for how resonant circuits operate. Displaying our waveforms, as shown in Figure 6-2:

vm(3)   is the magnitude response for Q = ½
vm(5)   is the magnitude response for Q = 1
vm(7)   is the magnitude response for Q = 2
vm(9)   is the magnitude response for Q = 4

We may display phases to show the phase response of the filter for different values of Q, as shown in Figure 6-3:

vp(3)   is the phase response for Q = ½
vp(5)   is the phase response for Q = 1
vp(7)   is the phase response for Q = 2
vp(9)   is the phase response for Q = 4

Since response is the ratio of output to input, you can look at the individual responses of a complicated filter chain this way. Probe will do the division for you,

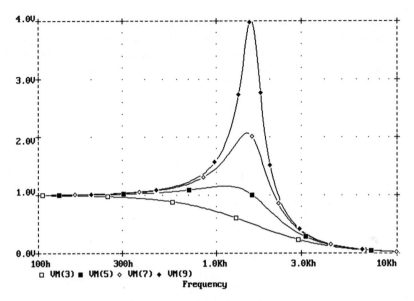

FIGURE 6-2 Plot of magnitude response.

which is a function you rarely find on an oscilloscope. Furthermore, if you were displaying response in decibels, you need only to subtract the decibel level of the input from the decibel level of the output. For example, if you had a two-stage filter with stage "A" feeding stage "B," the response of stage "B" is

$$\mathrm{db}(\mathrm{v}(\textit{B-node}))-\mathrm{db}(\mathrm{v}(\textit{A-node}))$$

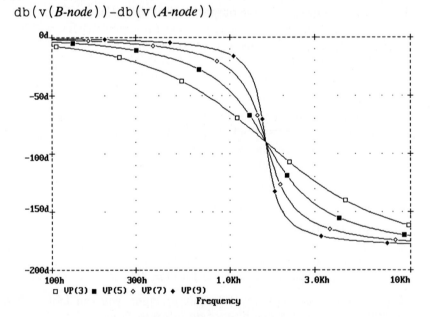

FIGURE 6-3 Plot of phase response.

A similar technique works for displaying phase angles. In this case, you merely subtract the input phase angle from the output phase angle. Using the previous example of the two-stage filter, the phase response of stage "B" is

$$\text{vp}(B\text{-}node) - \text{vp}(A\text{-}node)$$

6.5 PLOTTING GROUP DELAY

An important characteristic of a filter, or signal processing circuit, is its phase response, which is related to the distortion of the signal's waveform (shape) as it passes through the filter. (The phase response of the filter also demonstrates a link between frequency analysis and transient [time domain] analysis.) A circuit can maintain the wave-shape of a signal if the phase shift of each frequency component of the spectrum of the signal is a linear function of frequency. This means that if the circuit has uniform, or flat, delay, the signal is shifted only in time but is otherwise unchanged.

If the phase shift is a linear function of frequency, then its derivative with respect to frequency is constant. As it happens this same derivative

$$\frac{-\partial\,\text{phase}}{\partial\,\text{frequency}} \tag{6-1}$$

is also the delay time through the circuit for each frequency component (and, since we are talking about time **delay**, a minus sign is used so that time delays are stated as positive values). In the realm of modulated transmission, this is the delay time of the components in the envelope of a modulated carrier, so the delay is called *envelope delay*. In the realm of data transmission where the signals are pulses, this is the delay time for groups of pulses, so the delay is called *group delay*. In either case, the integrity of the transmission may depend on the variation in the delay time.

Probe will calculate group delay by using phase and frequency differences. A three-point calculation, using the adjacent data points (with special handling at the end points), is used to refine the derivative calculation. So long as the group delay is not changing very rapidly between each frequency increment in the analysis, the error of this technique will be negligible. We may display group delay to show the response of the filter for different values of Q, as shown in Figure 6-4:

vg(3)   is the group delay for Q = ½
vg(5)   is the group delay for Q = 1
vg(7)   is the group delay for Q = 2
vg(9)   is the group delay for Q = 4

EXERCISE 6.5-1

Develop the formula that Probe uses for calculating group delay for output voltage. (Hint: use the derivative of the phase of the output voltage.) Display both plots to check your formula.

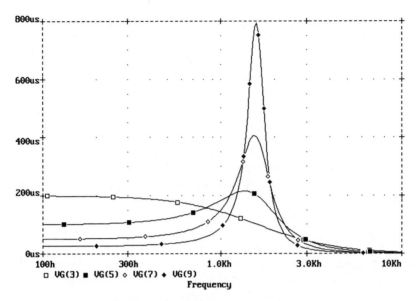

FIGURE 6-4 Plot of group delay.

6.6 USING MACROS: PLOTTING TIME DELAY

Whereas group delay is the amount of time required for the envelope of a modulated signal to pass through a filter, *time delay* is the delay for the carrier to pass through the same filter. This time is measured by the delay in zero crossings, assuming no DC bias. To say it another way, it is the time for any steady-state frequency to pass through the filter. Time delay is usually of more interest in audio frequency or signal processing circuits, while group delay is usually of more interest in communications circuits.

We can derive a formula for time delay that uses the phase response calculated by PSpice, by noting that the phase response value is a relative phase measurement. This means that the value displayed, which we think of as "phase," is actually $phase_{output}-phase_{input} = \Delta phase$ or "delta phase," although usually the input phase is zero. Now it is simple arithmetic:

$$\text{steady state frequency} = \frac{1}{360°} \frac{\Delta \text{phase}}{\Delta \text{time}}$$

$$\therefore \Delta \text{time} = \frac{1}{360°} \frac{\Delta \text{phase}}{\text{steady state frequency}}$$

(6-2)

where the "steady-state frequency" is simply the frequency of the analysis, and Δ time is "delta time." As with the formula for group delay, since we are talking about time **delay**, we will change the sign of the result so that time delays are stated as positive values.

When viewing time delay we will always be entering the formula derived in (6-2). It would be convenient to define a macro expansion that would do this for us. Probe has a "macro" facility, which is called from the menu, for defining and altering macro definitions. We can use macros to create new functions, or to use as a shorthand for commonly repeated waveform calculations. The macros are kept in a special definition file to save them for subsequent Probe sessions. The general form for a macro definition is

$$name[\![\,(arguments...\,)\,]\!] \;=\; formula$$

where the argument(s), if any, are replaced by name in the defining formula. Simple letters, such as x, y, etc., are normally used as argument names for brevity. However, longer names could be used to clarify the formula.

In this case, we would define the following macro

```
TD(p) = (-(p)/(360*frequency))
```

where "p" is the phase value to be evaluated, which could be the phase response of a node's voltage, a device's current, or any formula (such as the difference in phase between two stages in a filter). In the macro expansion, an extra set of parentheses is included around the use of the argument to assure proper calculation in the event a formula is used as the argument. Likewise in the macro expansion, an extra set of parentheses is included around the entire expansion to assure proper calculation in the event the macro is used in a formula or another macro definition.

Using the macro as a built-in function, we may display time delay to show the response of the filter for different values of Q, as shown in Figure 6-4, Figure 6-5:

```
td(vp(3))
```
 is the time delay for Q = ½
```
td(vp(5))
```
 is the time delay for Q = 1
```
td(vp(7))
```
 is the time delay for Q = 2
```
td(vp(9))
```
 is the time delay for Q = 4

It is instructive to note, once again, the difference between group delay and time delay: *group delay* is the delay in the modulating envelope of a steady-state carrier frequency; *time delay* is the delay in the zero crossings of that carrier. For our filter example, the difference between group delay and time delay can be seen in Figure 6-6, for the Q = ½ and Q = 4 cases.

6.7 COMPLEX VALUES

When using frequency analysis, all of the calculated voltages and currents are complex values. That is, these quantities are expressed as complex numbers, which have both **real** and **imaginary** components, and are not simply scalar values. You may treat (mathematically) the values in Probe the same way you would treat complex numbers (some of those things you learned in math class might be useful after all). Let's review the basics of manipulating complex values.

Suppose we have a complex number, "A," which is a combination of two real numbers, "a" and "b," by the formula

$$A = a + bi \tag{6-3}$$

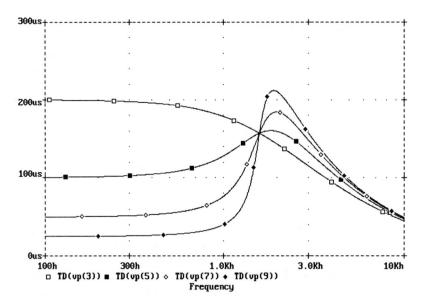

FIGURE 6-5 Plot of time delay, using a Probe macro.

where "i" is an imaginary quantity

$$i = \sqrt{-1} \tag{6-4}$$

The absolute value, or "modulus" (magnitude), of A is

$$|A| = \sqrt{a+b} \tag{6-5}$$

and the angle (phase) of A is

$$\angle A = tan^{-1}(b/a) \tag{6-6}$$

Using the magnitude and phase of A, we may re-specify a complex value as

$$A = a + bi = |A|e^{i\angle A} \tag{6-7}$$

where e is Euler's number (approximately 2.71828). This is another transformation available in complex math that we will take advantage of.

So, in Probe, when you call for the magnitude, or phase, of a voltage you are displaying the absolute value and angle of the complex quantity.

VM(5)   is the magnitude of the voltage at node 5
V(5)   is also the magnitude of the voltage at node 5, for convenience
VP(5)   is the phase of the voltage at node 5

If we have two complex values, A = a + bi and B = c + di, you may remember that the addition formula for complex numbers is

$$\begin{aligned} A + B &= (a + bi) + (c + di) \\ &= (a+c) + (b+d)i \end{aligned} \tag{6-8}$$

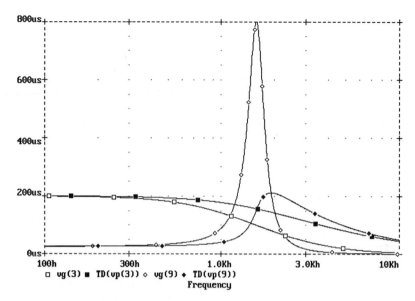

FIGURE 6-6 Plot of group delay and time delay.

This is the technique Probe uses when you add, or subtract, complex voltages or currents. You may also remember the formula for multiplication

$$
\begin{aligned}
A \cdot B &= (a + bi)(c + di) \\
&= (ac - bd) + (bc + ad)i \\
&= |A| \cdot |B| e^{i(\angle A + \angle B)}
\end{aligned}
\tag{6-9}
$$

and the formula for division

$$
\begin{aligned}
\frac{A}{B} &= \frac{a + bi}{c + di} \\
&= \frac{(ac + bd)}{(c^2 + d^2)} + \frac{(bc - ad)}{(c^2 + d^2)} i \\
&= \frac{|A|}{|B|} e^{i(\angle A - \angle B)}
\end{aligned}
\tag{6-10}
$$

These techniques are also in Probe for multiplying and dividing complex voltages or currents. All of these formulas become handy when you want the equations like V = IR, but V and I are complex quantities. The resistance R could also be a complex quantity, in which case it is usually called an impedance Z. We'll try this in the next section.

6.8 PLOTTING INPUT IMPEDANCE

Quite often, input impedance is an important circuit parameter. Usually it will vary with frequency. Without going into the problem of designing required impedance levels, let's look at how PSpice may be used to simulate and display input impedance.

Impedance is similar to resistance in that it is a coefficient in the observed change in voltage and current, so impedance is a complex resistance. From Ohm's law we know that $R = V \div I$. In a similar fashion, if V is a complex voltage

$$V = a + bi = |V| e^{i \angle V} \tag{6-11}$$

(where $|V|$ and $\angle V$ are the magnitude and phase angle of V) and I is a complex current

$$I = c + di = |I| e^{i \angle I} \tag{6-12}$$

(where $|I|$ and $\angle I$ are the magnitude and phase angle of I), and impedance is

$$
\begin{aligned}
Z = \frac{V}{I} &= \frac{a + bi}{c + di} \\
&= \frac{(ac + bd)}{(c^2 + d^2)} + \frac{(bc - ad)}{(c^2 + d^2)} i \\
&= \frac{|V|}{|I|} e^{i(\angle V - \angle I)} \\
&= |Z| e^{i \angle Z}
\end{aligned}
\tag{6-13}
$$

From this, we see that

$$
\begin{aligned}
|Z| &= \frac{|V|}{|I|} \\
\angle Z &= \angle V - \angle I
\end{aligned}
\tag{6-14}
$$

By using this result we may directly plot the magnitude, phase, real part, or imaginary part of input or output impedance. Using the same LC-filter circuit from before, we can plot input impedance without rerunning the simulation. All the data has already been calculated from the simulations for output response. The magnitude of the input impedance, or $|Z_{in}|$ as shown in Figure 6-7, may be plotted using

vm(1)/im(r1)	is $|Z_{in}|$ for Q = ½
vm(1)/im(r2)	is $|Z_{in}|$ for Q = 1
vm(1)/im(r3)	is $|Z_{in}|$ for Q = 2
vm(1)/im(r4)	is $|Z_{in}|$ for Q = 4

Note that vm(1) is the magnitude of the input voltage, which was set to a constant value that could have been used in the formula. Also, note that the Y-axis was set to a logarithmic scale. This has the same effect as displaying, say, output response in decibels, as in a *Bode plot*. You can see the "peaking" of the input impedance for Q > 1, and the asymptotic trend of impedance as the frequency tends toward low or high frequencies.

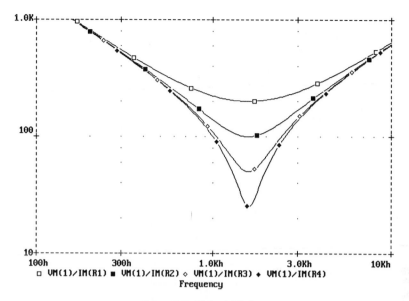

FIGURE 6-7 Plot of $|Z_{in}|$.

The phase angle of the input impedance, or $\angle Z_{in}$ as shown in Figure 6-8, may be plotted using

-ip(r1)	is $\angle Z_{in}$ for Q = ½
-ip(r2)	is $\angle Z_{in}$ for Q = 1
-ip(r3)	is $\angle Z_{in}$ for Q = 2
-ip(r4)	is $\angle Z_{in}$ for Q = 4

In this case, we "cheat" by knowing that the phase angle of the input voltage is zero.

EXERCISE 6.8-1

Develop a complete formula for the phase of an impedance. Use the real and imaginary parts of the voltage across the impedance and the current flowing through the impedance.

EXERCISE 6.8-2

Using the LC-filter circuit simulation, display Probe output for the following values: VI(1) and VI(2). What does this tell you about the input impedance? Now, after clearing the display, plot these values: I(R1), IR(R1), and II(R1). Where does the imaginary part of the input current cross the X-axis? What is significant about this frequency value? Try plotting the formula for the magnitude of the input current using the real and imaginary parts (only).

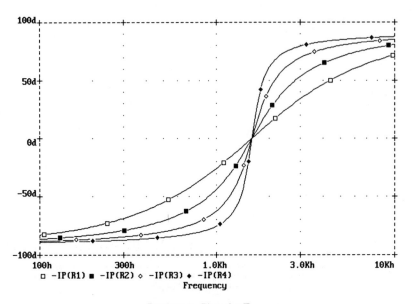

FIGURE 6-8   Plot of ∠Z_{in}.

From the preceding formulas for complex arithmetic, you can see that some thought is needed to use the real and imaginary components to calculate items like input impedance. So you will be happy to learn that **Probe does this for you**: for AC analysis, all of the arithmetic in Probe is done in complex form. This includes the scientific functions like SQRT( ), SIN( ), and LOG( ). The result of a calculation is displayed as a magnitude, by default, or you can use one of the following conversion functions:

> M( )   converts a complex value to its magnitude (you would normally not use this for displaying a response, but it might prove useful in converting a response into a value used to scale, say, a phase response)

> P( )   converts a complex value to its phase angle

> G( )   converts a complex value to its group delay, which you may recall is $-\partial$phase / ∂frequency

> R( )   converts complex value to its real part

> IMG( )   converts a complex value to its imaginary part

While there are no intrinsic functions for converting from (*magnitude,phase*) or (*real,imaginary*) components into complex values, you can define macros for these operations should you need them. Simply compose the required complex value; for example:

```
C2C(re,im) = ( re + im*sqrt(-1) )
```

will create a complex value from "Cartesian" components, and

```
P2C(mag,phs) = (mag*( cos(phs) + sin(phs)*sqrt(-1) ))
```

will create a complex value from "polar" components.

So for the previous example, instead of plotting $|Z_{in}|$ as `vm( )/im( )` you can use `M(v( )/i( ))`. Likewise, instead of plotting $\angle Z_{in}$ as `-ip(rx)`, and relying on the input phase being zero, you can use `P(v( )/i( ))`.

The real part of the input impedance, or $re(Z_{in})$ as shown in Figure 6-9, may be plotted using

`R(v(1)/i(r1))`	is $re(Z_{in})$ for Q = ½
`R(v(1)/i(r2))`	is $re(Z_{in})$ for Q = 1
`R(v(1)/i(r3))`	is $re(Z_{in})$ for Q = 2
`R(v(1)/i(r4))`	is $re(Z_{in})$ for Q = 4

Similarly, the imaginary part of the input impedance, or $im(Z_{in})$ as shown in Figure 6-10, may be plotted using

`IMG(v(1)/i(r1))`	is $im(Z_{in})$ for Q = ½
`IMG(v(1)/i(r2))`	is $im(Z_{in})$ for Q = 1
`IMG(v(1)/i(r3))`	is $im(Z_{in})$ for Q = 2
`IMG(v(1)/i(r4))`	is $im(Z_{in})$ for Q = 4

This takes all of the work out of complex displays.

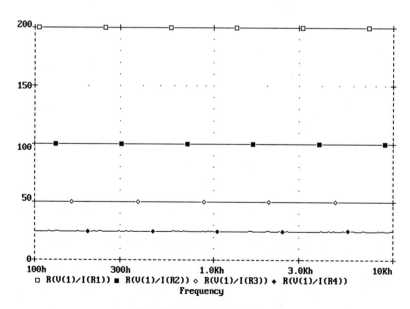

FIGURE 6-9 Plot of $re(Z_{in})$.

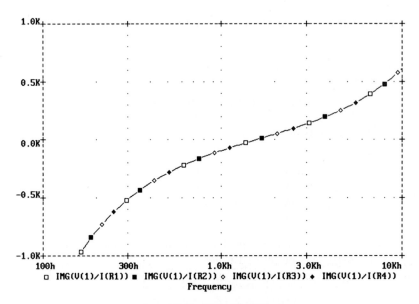

FIGURE 6-10   Plot of $im(Z_{in})$.

EXERCISE 6.8-3

Develop a complete formula for the imaginary part of the input impedance. As before, the magnitude and phase of the input voltage are unknown.

EXERCISE 6.8-4

You may have noticed that all of the plots for the imaginary part of the input impedance were the same. This means that the complex impedance part of the circuit was unchanged, regardless of which Q values we were investigating. Re-implement the Q = 4 circuits with real input impedance of 50 ohms and the same resonant frequency as the example. Then, plot the imaginary parts of the input impedance. Compare it to the example for Q = 2.

6.9 PLOTTING OUTPUT IMPEDANCE

Output impedance is, quite often, another important circuit parameter. We can use PSpice to simulate voltages and currents and then use Probe to display formulations for output impedance, just as we did previously for input impedance. You can even use these techniques for deriving the impedance levels internal to circuits, so long as you are careful to include the correct voltages and currents.

We were lucky, in the previous example, that one simulation run provided everything we needed for output response and a variety of ways of looking at input

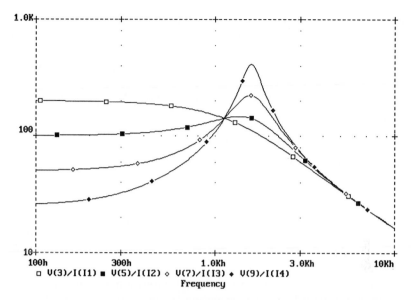

FIGURE 6-11   Plot of $|Z_{out}|$.

impedance. Applying a voltage source to the input was necessary to excite the circuit for simulating the output response, and also supplied current to the circuit from which the impedance calculations were done. To look at output impedance, we need to have current flowing at the output terminals to calculate impedance levels.

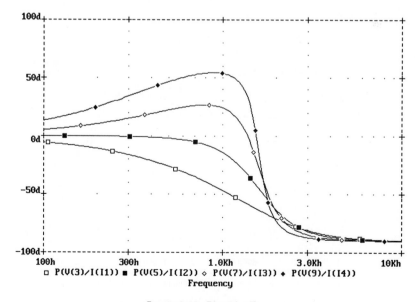

FIGURE 6-12   Plot of $\angle Z_{out}$.

For the LC-filter circuit, the input impedance calculations assumed that the output was not connected to a load. This is the same as saying the load had "infinite" impedance. The output impedance of the circuit assumes that the input is connected to a voltage source, which is to say that the input will have a load of zero impedance (the input is, for frequency calculations, "shorted"). To make the appropriate changes to our circuit file we will need to set the input voltage to zero:

```
Vin 1 0 0
```

and apply some current generators at the output of each filter section:

```
I1 0 3 AC 1
I2 0 5 AC 1
I3 0 7 AC 1
I4 0 9 AC 1
```

Note that the current sources are connected so that positive current flows into the output node of the LC-filter. This is done so the phasing of the current will be intuitively correct when we display the output impedance. Now, we rerun the simulation.

Knowing that Probe performs complex arithmetic simplifies the task of plotting the output impedance. The magnitude of the output impedance, or $|Z_{out}|$ as shown in Figure 6-11, may be plotted using

$$v(3)/i(11) \quad \text{is } |Z_{out}| \text{ for } Q = \tfrac{1}{2}$$
$$v(5)/i(12) \quad \text{is } |Z_{out}| \text{ for } Q = 1$$
$$v(7)/i(13) \quad \text{is } |Z_{out}| \text{ for } Q = 2$$
$$v(9)/i(14) \quad \text{is } |Z_{out}| \text{ for } Q = 4$$

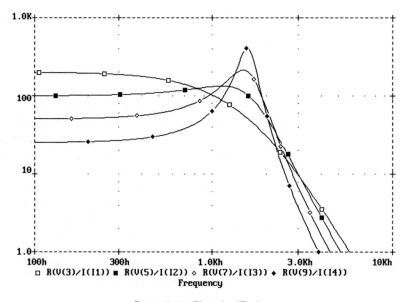

FIGURE 6-13 Plot of $re(Z_{out})$.

Note that i(11), i(12), i(13), and i(14) represent the magnitude of the driving current, which was set to a constant value that could have been used in the formula. Also, note that the Y-axis was set to a logarithmic scale. This has the same effect as displaying, say, output response in decibels. You can see the "peaking" of the output impedance for Q > 1, and the asymptotic trend of impedance as the frequency tends toward low or high frequencies.

The phase angle of the output impedance, or $\angle Z_{out}$ as shown in Figure 6-12, may be plotted using

P(v(3)/i(11))	is $\angle Z_{out}$ for Q = ½
P(v(5)/i(12))	is $\angle Z_{out}$ for Q = 1
P(v(7)/i(13))	is $\angle Z_{out}$ for Q = 2
P(v(9)/i(14))	is $\angle Z_{out}$ for Q = 4

The real part of the output impedance, or $re(Z_{out})$ as shown in Figure 6-13, may be plotted using

R(v(3)/i(11))	is $re(Z_{out})$ for Q = ½
R(v(5)/i(12))	is $re(Z_{out})$ for Q = 1
R(v(7)/i(13))	is $re(Z_{out})$ for Q = 2
R(v(9)/i(14))	is $re(Z_{out})$ for Q = 4

Again we "cheat" by knowing that the driving current source has no imaginary part, which is another way of saying that its phase angle is zero.

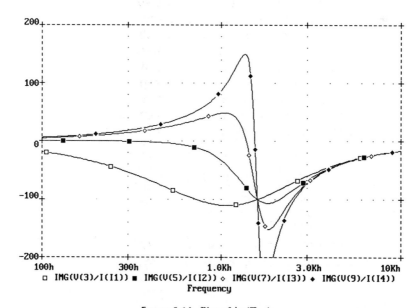

FIGURE 6-14 Plot of $im(Z_{out})$.

EXERCISE 6.9-1

> Develop a complete formula for the real part of the output impedance. In this case, the magnitude and phase of the output voltage are unknown.

The imaginary part of the output impedance, or $im(Z_{out})$ as shown in Figure 6-14, may be plotted using

> IMG(v(3)/i(11))   is $im(Z_{out})$ for Q = ½
> IMG(v(5)/i(12))   is $im(Z_{out})$ for Q = 1
> IMG(v(7)/i(13))   is $im(Z_{out})$ for Q = 2
> IMG(v(9)/i(14))   is $im(Z_{out})$ for Q = 4

Again, we "cheat" by knowing that the driving current source has no imaginary part. Also, we take advantage of the real part of the driving current being unity.

EXERCISE 6.9-2

> Develop a complete formula for the imaginary part of the output impedance. As before, the magnitude and phase of the output voltage are unknown.

CHAPTER 7

TWO-PORT NETWORK ANALYSIS

While linear network theory covers multiport networks, there is considerable emphasis on *two-port networks* as an important class of general circuit. Active devices in practical circuits are mainly two-port networks, with an *input port* and an *output port*, and the transmission of signals from input to output is governed by the network parameters. Many passive networks, which are only composed of linear time-invariant resistors, capacitors, and inductors, are also classified as two-port networks. Active networks provide power gain and have the potential for oscillation. Passive networks are inherently stable and bidirectional. For both types of network, the engineer is often interested in designing optimum terminations, for the ports, to maximize power transfer through the network. Two-port representations are a complete and simple way of describing a network while suppressing all detail of internal structure.

A generalized diagram for a two-port network is shown in Figure 7-1. The port variables are voltage and current (note that current, by convention, is always measured **into** the port of interest). The various two-port parameter sets are matrix elements that define the linear relationship between the port variables. Different two-port parameter sets are more useful and appropriate for different design problems. For example, scattering parameters may exist for passive networks, such as transformers, which do not have admittance or impedance parameters. Similarly, some active devices, such as triode and pentode vacuum tubes, have admittance and hybrid parameters but do not have impedance parameters.

PSpice, with its origins in integrated circuit analysis, would not seem appropriate for this style of linear circuit analysis. However, a little study of what these two-port parameter sets represent will provide an avenue for analyzing circuits this way.

FIGURE 7-1 Two-port active circuit.

7.1 OPEN-CIRCUIT IMPEDANCE PARAMETERS $[z_{ij}]$

Looking at the generalized diagram in Figure 7-1, if we take I_1 and I_2 as the independent values, then V_1 and V_2 can be given by

$$V_1 = z_{11} \cdot I_1 + z_{12} \cdot I_2$$
$$V_2 = z_{21} \cdot I_1 + z_{22} \cdot I_2$$

(7-1)

or in a matrix form

$$\begin{bmatrix} V_1 \\ V_2 \end{bmatrix} = \begin{bmatrix} z_{11} & z_{12} \\ z_{21} & z_{22} \end{bmatrix} \begin{bmatrix} I_1 \\ I_2 \end{bmatrix}$$

(7-2)

The coefficients z_{ij} have the dimensions of *impedance*. As you can see from the Thévenin equivalent circuit, in Figure 7-2, these parameters can be easily measured by test arrangements wherein one of the ports is left as an open circuit. This is how their name *open-circuit impedance* parameters comes about. For example, you would naturally measure z_{11} by leaving the second port as an open circuit. This would force the current I_2 to be zero, and then the only influence on V_1 would be I_1. So, z_{11} is the ratio of V_1 to I_1. In fact, z_{ij} is the ratio of V_i to I_j where the other current (I_i) is zero, or for all of the ratios

$$z_{11} = \left. \frac{V_1}{I_1} \right|_{I_2 = 0} = \begin{array}{l} \textit{open-circuit} \\ \textit{input} \\ \textit{impedance} \end{array} \qquad z_{12} = \left. \frac{V_1}{I_2} \right|_{I_1 = 0} = \begin{array}{l} \textit{open-circuit} \\ \textit{reverse transfer} \\ \textit{impedance} \end{array}$$

(7-3)

$$z_{21} = \left. \frac{V_2}{I_1} \right|_{I_2 = 0} = \begin{array}{l} \textit{open-circuit} \\ \textit{forward transfer} \\ \textit{impedance} \end{array} \qquad z_{22} = \left. \frac{V_2}{I_2} \right|_{I_1 = 0} = \begin{array}{l} \textit{open-circuit} \\ \textit{output} \\ \textit{impedance} \end{array}$$

Let's try measuring these parameters with PSpice. Note that while two-port parameters are generalized for small-signal analysis, the same matrix values apply for DC analysis of passive, non-reactive circuits. This means that you might find it easier to think about the simulated measurements by imagining an all-resistive circuit, with DC voltages and currents.

Since, in general, these parameters are ratios of small-signal voltages and currents, we will need to use a large-valued inductor to block any AC current at a port without upsetting the DC operating point of the circuit. So, our measuring circuits

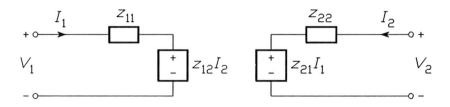

FIGURE 7-2 Thévenin equivalent circuit.

will isolate the device under test (DUT) and prevent AC current from affecting the bias circuity. Then we apply an AC current signal to each port to measure two matrix values from each set-up. This will require two set-ups to obtain all four matrix terms. For our example, the following circuit file for the two set-ups is

```
* measure z11 and z21
x1 11 21 TEST
i1 0  11 AC 1

* measure z12 and z22
x2 12 22 TEST
i2 0  22 AC 1

.print ac vm(11) vp(11) vm(12) vp(12)
.print ac vm(21) vp(21) vm(22) vp(22)
```

The .PRINT statements print the magnitude and phase of the port voltages, which have the same values as their respective matrix parameters, since the stimulating current is unity. In our example, the TEST circuit has the following definition:

```
.subckt TEST port1 port2
  vcc vc 0 DC 6
  rl  vc 2 50
  ll  2  port2 1 ; one Henry
  rs  0  1 50
  ls  1  port1 1 ; one Henry
  xd1 port1 port2 3 DUT
  ibias  3 0 DC 5m
  cblck  3 0 1   ; one Farad
.ends
```

which provides the proper DC bias and AC isolation. Finally, the DUT (device under test) circuit has the following definition:

```
.subckt DUT 1 2 3
  lb 1 b 2n
  le 3 e 2n
  q1 2 b e rf_npn
.model rf_npn NPN(Is=47.75E-18 Xti=3 Eg=1.11 Vaf=150 Bf=87.27
+ Ise=602.4f Ne=2.081 Ikf=87.9m Nk=.518 Xtb=1.5 Br=1 Isc=66.18E-18
+ Nc=2.171 Ikr=2.608 Rc=4.056 Cjc=1.014p Mjc=.1971 Vjc=.359 Fc=.5
+ Cje=1.425p Mje=.2725 Vje=.75 Tr=10n Tf=126.1p Itf=2.711
+ Xtf=372.6K Vtf=1.6 Rb=10 Ptf=40)
.ends
```

which includes an active device, a radio-frequency NPN transistor including its model definition, and two parasitic inductors that are part of the transistor's package. Realizing that we have not covered active devices and their models, you will need to take it, on faith, that this circuit is a representative model for a high-frequency transistor. The TEST circuit, with the DUT circuit embedded, is shown in Figure 7-3.

Combining these elements with a .AC statement for a 100MHz to 1GHz sweep, we run PSpice and inspect the output file for the .PRINT results. First, we have the

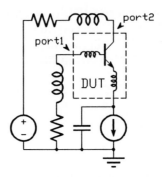

FIGURE 7-3 Schematic of TEST subcircuit.

magnitudes and phases for z_{11} and z_{12}:

FREQ	VM(11)	VP(11)	VM(12)	VP(12)
1.000E+08	1.525E+01	6.773E+00	5.473E+00	1.032E+01
2.000E+08	1.572E+01	1.574E+01	5.708E+00	1.999E+01
3.000E+08	1.646E+01	2.355E+01	6.078E+00	2.874E+01
4.000E+08	1.745E+01	3.047E+01	6.556E+00	3.639E+01
5.000E+08	1.864E+01	3.657E+01	7.119E+00	4.300E+01
6.000E+08	2.000E+01	4.190E+01	7.746E+00	4.868E+01
7.000E+08	2.148E+01	4.655E+01	8.421E+00	5.361E+01
8.000E+08	2.307E+01	5.064E+01	9.137E+00	5.793E+01
9.000E+08	2.475E+01	5.424E+01	9.888E+00	6.178E+01
1.000E+09	2.650E+01	5.744E+01	1.067E+01	6.527E+01

Then come the magnitudes and phases for z_{21} and z_{22}:

FREQ	VM(21)	VP(21)	VM(22)	VP(22)
1.000E+08	2.579E+03	8.378E+01	3.027E+02	−8.815E+00
2.000E+08	1.292E+03	7.748E+01	3.013E+02	−8.511E+00
3.000E+08	8.644E+02	7.119E+01	3.008E+02	−1.027E+01
4.000E+08	6.510E+02	6.491E+01	3.004E+02	−1.261E+01
5.000E+08	5.231E+02	5.862E+01	2.998E+02	−1.522E+01
6.000E+08	4.377E+02	5.233E+01	2.988E+02	−1.804E+01
7.000E+08	3.763E+02	4.603E+01	2.975E+02	−2.102E+01
8.000E+08	3.297E+02	3.972E+01	2.955E+02	−2.415E+01
9.000E+08	2.927E+02	3.341E+01	2.929E+02	−2.743E+01
1.000E+09	2.621E+02	2.710E+01	2.894E+02	−3.084E+01

Of course, these values could also be displayed graphically with Probe, but it is easier to inspect these numbers for comparing the matrix parameter values at any given frequency.

7.2 SHORT-CIRCUIT ADMITTANCE PARAMETERS [Y$_{ij}$]

Looking back at the generalized diagram in Figure 7-1, if we take V_1 and V_2 as the independent values, then I_1 and I_2 can be given by

$$I_1 = y_{11} \cdot V_1 + y_{12} \cdot V_2$$
$$I_2 = y_{21} \cdot V_1 + y_{22} \cdot V_2$$

(7-4)

or in a matrix form

$$\begin{bmatrix} I_1 \\ I_2 \end{bmatrix} = \begin{bmatrix} y_{11} & y_{12} \\ y_{21} & y_{22} \end{bmatrix} \begin{bmatrix} V_1 \\ V_2 \end{bmatrix}$$

(7-5)

The coefficients y_{ij} have the dimensions of *admittance*. As you can see from the Norton equivalent circuit, in Figure 7-4, these parameters can be easily measured by test arrangements wherein one of the ports is short circuited. This is how their name *short-circuit admittance* parameters comes about. For example, you would naturally measure y_{11} by shorting the second port. This would force the voltage V_2 to be zero, and then the only influence on I_1 would be V_1. So, y_{11} is the ratio of I_1 to V_1. In fact, y_{ij} is the ratio of I_i to V_j where the other voltage (V_i) is zero, or for all of the ratios

$$y_{11} = \left. \frac{I_1}{V_1} \right|_{V_2 = 0} = \begin{array}{l} \textit{short-circuit} \\ \textit{input} \\ \textit{admittance} \end{array} \qquad y_{12} = \left. \frac{I_1}{V_2} \right|_{V_1 = 0} = \begin{array}{l} \textit{short-circuit} \\ \textit{reverse transfer} \\ \textit{admittance} \end{array}$$

(7-6)

$$y_{21} = \left. \frac{I_2}{V_1} \right|_{V_2 = 0} = \begin{array}{l} \textit{short-circuit} \\ \textit{forward transfer} \\ \textit{admittance} \end{array} \qquad y_{22} = \left. \frac{I_2}{V_2} \right|_{V_1 = 0} = \begin{array}{l} \textit{short-circuit} \\ \textit{output} \\ \textit{admittance} \end{array}$$

Let's try measuring these parameters with PSpice. As mentioned before, note that while two-port parameters are generalized for small-signal analysis, the same matrix values apply for DC analysis of passive, non-reactive circuits. This means that you might find it easier to think about the simulated measurements by imagining an all-resistive circuit, with DC voltages and currents.

As in the previous section on z_{ij} parameters, our measuring circuits will isolate the device under test (DUT) and use large-valued DC-blocking (that is, AC-conducting) capacitors to enforce the short-circuit condition. Then we apply an AC voltage signal, through large-valued capacitors, to each port to measure two matrix values from each

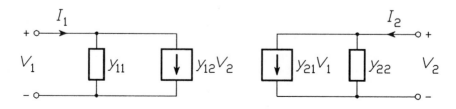

FIGURE 7-4 Norton equivalent circuit.

set-up. This will require two set-ups to obtain all four matrix terms. For our example, the circuit file for the two set-ups is

```
* measure y11 and y21
x1   11  21  TEST
v11  11  a1  AC 1
c11  0   a1  1 ; one Farad
v21  21  a2  0
c21  0   a2  1 ; one Farad

* measure y12 and y22
x2   12  22  TEST
v12  12  b1  0
c12  0   b1  1 ; one Farad
v22  22  b2  AC 1
c22  0   b2  1 ; one Farad

.print ac im(c11) ip(c11) im(c12) ip(c12)
.print ac im(c21) ip(c21) im(c22) ip(c22)
```

The .PRINT statements print the magnitude and phase of current into the ports, which have the same values as their respective matrix parameters, since the stimulating voltage is unity. Since the value printed is for the current through the capacitor, it is important to connect the capacitors to correctly show the phase of the measured current. The TEST and DUT (device under test) circuits have the same definitions as shown on page 69, in the section on z_{ij} parameters.

Combining these elements with a .AC statement for a 100MHz to 1GHz sweep, we run PSpice and inspect the output file for the .PRINT results. First, we have the magnitudes and phases for y_{11} and y_{12}:

FREQ	IM(c11)	IP(c11)	IM(c12)	IP(c12)
1.000E+08	1.976E-02	5.964E+01	3.574E-04	-1.012E+02
2.000E+08	3.430E-02	4.139E+01	6.499E-04	-1.101E+02
3.000E+08	4.294E-02	2.491E+01	8.674E-04	-1.161E+02
4.000E+08	4.715E-02	1.125E+01	1.029E-03	-1.198E+02
5.000E+08	4.858E-02	-6.746E-02	1.154E-03	-1.218E+02
6.000E+08	4.835E-02	-9.565E+00	1.253E-03	-1.228E+02
7.000E+08	4.716E-02	-1.766E+01	1.335E-03	-1.230E+02
8.000E+08	4.544E-02	-2.464E+01	1.405E-03	-1.226E+02
9.000E+08	4.344E-02	-3.072E+01	1.467E-03	-1.215E+02
1.000E+09	4.132E-02	-3.608E+01	1.525E-03	-1.200E+02

Then come the magnitudes and phases for y_{21} and y_{22}:

FREQ	IM(c21)	IP(c21)	IM(c22)	IP(c22)
1.000E+08	1.684E-01	-2.776E+01	9.959E-04	7.523E+01
2.000E+08	1.472E-01	-5.262E+01	1.789E-03	6.565E+01
3.000E+08	1.234E-01	-7.362E+01	2.349E-03	5.873E+01
4.000E+08	1.022E-01	-9.124E+01	2.740E-03	5.433E+01
5.000E+08	8.478E-02	-1.062E+02	3.022E-03	5.173E+01
6.000E+08	7.082E-02	-1.192E+02	3.236E-03	5.037E+01
7.000E+08	5.967E-02	-1.306E+02	3.406E-03	4.991E+01
8.000E+08	5.069E-02	-1.408E+02	3.548E-03	5.016E+01
9.000E+08	4.341E-02	-1.499E+02	3.672E-03	5.096E+01
1.000E+09	3.743E-02	-1.581E+02	3.785E-03	5.220E+01

As mentioned before, these values could also be displayed graphically with Probe, but it is easier to inspect these numbers for comparing the matrix parameter values at any given frequency.

7.3 HYBRID PARAMETERS [H_{ij}]

Looking back at the generalized diagram in Figure 7-1, if we take I_1 and V_2 as the independent values, then V_1 and I_2 can be given by

$$V_1 = h_{11} \cdot I_1 + h_{12} \cdot V_2$$
$$I_2 = h_{21} \cdot I_1 + h_{22} \cdot V_2$$

(7-7)

or in a matrix form

$$\begin{bmatrix} V_1 \\ I_2 \end{bmatrix} = \begin{bmatrix} h_{11} & h_{12} \\ h_{21} & h_{22} \end{bmatrix} \begin{bmatrix} I_1 \\ V_2 \end{bmatrix}$$

(7-8)

Unlike the impedance or admittance matrix, the coefficients h_{ij} have various dimensions. As you can see from the equivalent circuit, in Figure 7-5, these parameters can be easily measured by test arrangements that are a combination of the techniques used for the impedance and admittance parameters. For all of the ratios

$$h_{11} = \frac{V_1}{I_1}\bigg|_{V_2=0} = \begin{array}{l}\textit{short-circuit}\\\textit{input}\\\textit{impedance}\end{array} \qquad h_{12} = \frac{V_1}{V_2}\bigg|_{I_1=0} = \begin{array}{l}\textit{open-circuit}\\\textit{reverse}\\\textit{voltage gain}\end{array}$$

$$h_{21} = \frac{I_2}{I_1}\bigg|_{V_2=0} = \begin{array}{l}\textit{short-circuit}\\\textit{forward}\\\textit{current gain}\end{array} \qquad h_{22} = \frac{I_2}{V_2}\bigg|_{I_1=0} = \begin{array}{l}\textit{open-circuit}\\\textit{output}\\\textit{admittance}\end{array}$$

(7-9)

Note that each coefficient has different units: this is how their name *hybrid* parameters comes about.

Let's try measuring these parameters with PSpice. As mentioned before, note that while two-port parameters are generalized for small-signal analysis, the same matrix values apply for DC analysis of passive, non-reactive circuits. This means that you might find it easier to think about the simulated measurements by imagining an all-resistive circuit, with DC voltages and currents.

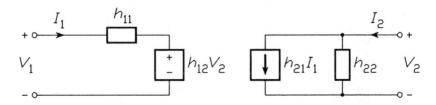

FIGURE 7-5 Equivalent circuit.

Even though we will use a hybrid of the techniques shown in the previous sections on z_{ij} and y_{ij} parameters, there are still only two port conditions to enforce. This will require two set-ups to obtain all four matrix terms. For our example, the circuit file for the two set-ups is

```
* measure h11 and h21
x1   11 21  TEST
i1   0   11  AC 1
v21 21 a2 0
c21 0   a2 1 ; one Farad

* measure h12 and h22
x2   12 22  TEST
v22 22 b2 AC 1
c22 0   b2 1 ; one Farad

.print ac vm( 11) vp( 11) vm( 12) vp( 12)
.print ac im(c21) ip(c21) im(c22) ip(c22)
```

The .PRINT statements print the magnitude and phase of port voltages and port currents, which have the same values as their respective matrix parameters, since the stimulating voltage or current is unity. When the value printed is for the current through the capacitor, it is important to connect the capacitors to correctly show the phase of the measured current. The TEST and DUT (device under test) circuits have the same definitions as shown on page 69, in the section on z_{ij} parameters.

Combining these elements with a .AC statement for a 100MHz to 1GHz sweep, we run PSpice and inspect the output file for the .PRINT results. First, we have the magnitudes and phases for h_{11} and h_{12}:

FREQ	VM(11)	VP(11)	VM(12)	VP(12)
1.000E+08	5.060E+01	-5.964E+01	1.808E-02	1.914E+01
2.000E+08	2.915E+01	-4.139E+01	1.895E-02	2.851E+01
3.000E+08	2.329E+01	-2.491E+01	2.020E-02	3.901E+01
4.000E+08	2.121E+01	-1.125E+01	2.182E-02	4.900E+01
5.000E+08	2.058E+01	6.759E-02	2.375E-02	5.822E+01
6.000E+08	2.068E+01	9.565E+00	2.592E-02	6.672E+01
7.000E+08	2.120E+01	1.765E+01	2.831E-02	7.462E+01
8.000E+08	2.201E+01	2.463E+01	3.092E-02	8.208E+01
9.000E+08	2.302E+01	3.072E+01	3.376E-02	8.921E+01
1.000E+09	2.420E+01	3.608E+01	3.689E-02	9.610E+01

Then come the magnitudes and phases for h_{21} and h_{22}:

FREQ	IM(c21)	IP(c21)	IM(c22)	IP(c22)
1.000E+08	8.521E+00	-8.740E+01	3.304E-03	8.814E+00
2.000E+08	4.290E+00	-9.401E+01	3.319E-03	8.513E+00
3.000E+08	2.873E+00	-9.854E+01	3.324E-03	1.027E+01
4.000E+08	2.167E+00	-1.025E+02	3.329E-03	1.260E+01
5.000E+08	1.745E+00	-1.062E+02	3.336E-03	1.523E+01
6.000E+08	1.465E+00	-1.096E+02	3.346E-03	1.803E+01
7.000E+08	1.265E+00	-1.130E+02	3.362E-03	2.102E+01
8.000E+08	1.116E+00	-1.161E+02	3.384E-03	2.415E+01
9.000E+08	9.993E-01	-1.192E+02	3.415E-03	2.743E+01
1.000E+09	9.056E-01	-1.221E+02	3.456E-03	3.084E+01

As mentioned before, these values could also be displayed graphically with Probe, but it is easier to inspect these numbers for comparing the matrix parameter values at any given frequency.

You may have seen these parameters before. Since the two-port network we were measuring is a bipolar transistor, in its common-emitter configuration, the *input*, *output*, *forward*, and *reverse* parameters are also called h_{ie}, h_{oe}, h_{fe}, and h_{re}, respectively. The first subscript denotes the parameter type, while the second subscript denotes the transistor's operating mode. For example, h_{fe} stands for "forward current-gain in a common-emitter configuration."

7.4 OTHER PARAMETER SETS

As you can see from the previous sections, the two-port parameters come from applying two sources, independently, at either (or both) ports, and measuring two responses. In other words, we have a linear system of four variables where two can be specified independently. The number of combinations available is

$$C_2^4 = \begin{pmatrix} 4 \\ 2 \end{pmatrix} = \frac{4!}{(4-2)! \cdot 2!} = 6 \tag{7-10}$$

So, there are only six different representations for the two-port system. Three of these were described in the previous sections, the others are

- the *inverse hybrid* parameters $[g_{ij}]$, which are the inverse of the hybrid parameters ($[g] = [h]^{-1}$)

$$\begin{bmatrix} I_1 \\ V_2 \end{bmatrix} = \begin{bmatrix} g_{11} & g_{12} \\ g_{21} & g_{22} \end{bmatrix} \begin{bmatrix} V_1 \\ I_2 \end{bmatrix} \tag{7-11}$$

- the *chain* parameters [*ABCD*], referred to as general circuit constants, or "ABCD" parameters (also sometimes called the "transmission" parameters).

$$\begin{bmatrix} V_1 \\ I_1 \end{bmatrix} = \begin{bmatrix} A & B \\ C & D \end{bmatrix} \begin{bmatrix} V_2 \\ -I_2 \end{bmatrix} \tag{7-12}$$

- the *inverse chain* parameters [*abcd*], also referred to as general circuit constants, or the "backward ABCD" parameters.

$$\begin{bmatrix} V_2 \\ I_2 \end{bmatrix} = \begin{bmatrix} a & b \\ c & d \end{bmatrix} \begin{bmatrix} V_1 \\ -I_1 \end{bmatrix} \tag{7-13}$$

Equations (7-12) and (7-13) are useful for analyzing of cascaded two-port networks, and for this purpose a negative sign is associated with one of the independent variables; the overall matrix of the cascaded network is simply the matrix multiplication of the constituent matrices.

TABLE 7-1 Two-port parameter conversions.

	$[z_{ij}]$	$[y_{ij}]$	$[h_{ij}]$	$[g_{ij}]$	$[A]$	$[a]$
$[z_{ij}]$	$\begin{matrix} z_{11} & z_{12} \\ z_{21} & z_{22} \end{matrix}$	$\begin{matrix} \dfrac{y_{22}}{\Delta_y} & \dfrac{-y_{12}}{\Delta_y} \\[6pt] \dfrac{-y_{21}}{\Delta_y} & \dfrac{y_{11}}{\Delta_y} \end{matrix}$	$\begin{matrix} \dfrac{\Delta_h}{h_{22}} & \dfrac{h_{12}}{h_{22}} \\[6pt] \dfrac{-h_{21}}{h_{22}} & \dfrac{1}{h_{22}} \end{matrix}$	$\begin{matrix} \dfrac{1}{g_{11}} & \dfrac{-g_{12}}{g_{11}} \\[6pt] \dfrac{g_{21}}{g_{11}} & \dfrac{\Delta_g}{g_{11}} \end{matrix}$	$\begin{matrix} \dfrac{A}{C} & \dfrac{\Delta_A}{C} \\[6pt] \dfrac{1}{C} & \dfrac{D}{C} \end{matrix}$	$\begin{matrix} \dfrac{d}{c} & \dfrac{1}{c} \\[6pt] \dfrac{\Delta_a}{c} & \dfrac{a}{c} \end{matrix}$
$[y_{ij}]$	$\begin{matrix} \dfrac{z_{22}}{\Delta_z} & \dfrac{-z_{12}}{\Delta_z} \\[6pt] \dfrac{-z_{21}}{\Delta_z} & \dfrac{z_{11}}{\Delta_z} \end{matrix}$	$\begin{matrix} y_{11} & y_{12} \\ y_{21} & y_{22} \end{matrix}$	$\begin{matrix} \dfrac{1}{h_{11}} & \dfrac{-h_{12}}{h_{11}} \\[6pt] \dfrac{h_{21}}{h_{11}} & \dfrac{\Delta_h}{h_{11}} \end{matrix}$	$\begin{matrix} \dfrac{\Delta_g}{g_{22}} & \dfrac{g_{12}}{g_{22}} \\[6pt] \dfrac{-g_{21}}{g_{22}} & \dfrac{1}{g_{22}} \end{matrix}$	$\begin{matrix} \dfrac{D}{B} & \dfrac{\Delta_A}{-B} \\[6pt] \dfrac{1}{-B} & \dfrac{A}{B} \end{matrix}$	$\begin{matrix} \dfrac{a}{b} & \dfrac{1}{-b} \\[6pt] \dfrac{\Delta_a}{-b} & \dfrac{d}{b} \end{matrix}$
$[h_{ij}]$	$\begin{matrix} \dfrac{\Delta_z}{z_{22}} & \dfrac{z_{12}}{z_{22}} \\[6pt] \dfrac{-z_{21}}{z_{22}} & \dfrac{1}{z_{22}} \end{matrix}$	$\begin{matrix} \dfrac{1}{y_{11}} & \dfrac{-y_{12}}{y_{11}} \\[6pt] \dfrac{y_{21}}{y_{11}} & \dfrac{\Delta_y}{y_{11}} \end{matrix}$	$\begin{matrix} h_{11} & h_{12} \\ h_{21} & h_{22} \end{matrix}$	$\begin{matrix} \dfrac{g_{22}}{\Delta_g} & \dfrac{-g_{12}}{\Delta_g} \\[6pt] \dfrac{-g_{21}}{\Delta_g} & \dfrac{g_{11}}{\Delta_g} \end{matrix}$	$\begin{matrix} \dfrac{B}{D} & \dfrac{\Delta_A}{D} \\[6pt] \dfrac{1}{-D} & \dfrac{C}{D} \end{matrix}$	$\begin{matrix} \dfrac{b}{a} & \dfrac{1}{a} \\[6pt] \dfrac{\Delta_a}{-a} & \dfrac{c}{a} \end{matrix}$
$[g_{ij}]$	$\begin{matrix} \dfrac{1}{z_{11}} & \dfrac{-z_{12}}{z_{11}} \\[6pt] \dfrac{z_{21}}{z_{11}} & \dfrac{\Delta_z}{z_{11}} \end{matrix}$	$\begin{matrix} \dfrac{\Delta_y}{y_{22}} & \dfrac{y_{12}}{y_{22}} \\[6pt] \dfrac{-y_{21}}{y_{22}} & \dfrac{1}{y_{22}} \end{matrix}$	$\begin{matrix} \dfrac{h_{22}}{\Delta_h} & \dfrac{-h_{12}}{\Delta_h} \\[6pt] \dfrac{-h_{21}}{\Delta_h} & \dfrac{h_{11}}{\Delta_h} \end{matrix}$	$\begin{matrix} g_{11} & g_{12} \\ g_{21} & g_{22} \end{matrix}$	$\begin{matrix} \dfrac{C}{A} & \dfrac{\Delta_A}{-A} \\[6pt] \dfrac{1}{A} & \dfrac{B}{A} \end{matrix}$	$\begin{matrix} \dfrac{c}{d} & \dfrac{1}{-d} \\[6pt] \dfrac{\Delta_a}{d} & \dfrac{b}{d} \end{matrix}$
$[A]$	$\begin{matrix} \dfrac{z_{11}}{z_{21}} & \dfrac{\Delta_z}{z_{21}} \\[6pt] \dfrac{1}{z_{21}} & \dfrac{z_{22}}{z_{21}} \end{matrix}$	$\begin{matrix} \dfrac{y_{22}}{-y_{21}} & \dfrac{1}{-y_{21}} \\[6pt] \dfrac{\Delta_y}{-y_{21}} & \dfrac{y_{11}}{-y_{21}} \end{matrix}$	$\begin{matrix} \dfrac{\Delta_h}{-h_{21}} & \dfrac{h_{11}}{-h_{21}} \\[6pt] \dfrac{h_{22}}{-h_{21}} & \dfrac{1}{-h_{21}} \end{matrix}$	$\begin{matrix} \dfrac{1}{g_{21}} & \dfrac{g_{22}}{g_{21}} \\[6pt] \dfrac{g_{11}}{g_{21}} & \dfrac{\Delta_g}{g_{21}} \end{matrix}$	$\begin{matrix} A & B \\ C & D \end{matrix}$	$\begin{matrix} \dfrac{d}{\Delta_a} & \dfrac{b}{\Delta_a} \\[6pt] \dfrac{c}{\Delta_a} & \dfrac{a}{\Delta_a} \end{matrix}$
$[a]$	$\begin{matrix} \dfrac{z_{11}}{z_{12}} & \dfrac{\Delta_z}{z_{12}} \\[6pt] \dfrac{1}{z_{12}} & \dfrac{z_{11}}{z_{12}} \end{matrix}$	$\begin{matrix} \dfrac{y_{11}}{-y_{12}} & \dfrac{1}{-y_{12}} \\[6pt] \dfrac{\Delta_y}{-y_{12}} & \dfrac{y_{22}}{-y_{12}} \end{matrix}$	$\begin{matrix} \dfrac{1}{h_{12}} & \dfrac{h_{11}}{h_{12}} \\[6pt] \dfrac{h_{22}}{h_{12}} & \dfrac{\Delta_h}{h_{12}} \end{matrix}$	$\begin{matrix} \dfrac{\Delta_g}{-g_{12}} & \dfrac{g_{22}}{-g_{12}} \\[6pt] \dfrac{g_{11}}{-g_{12}} & \dfrac{1}{-g_{12}} \end{matrix}$	$\begin{matrix} \dfrac{D}{\Delta_A} & \dfrac{B}{\Delta_A} \\[6pt] \dfrac{C}{\Delta_A} & \dfrac{A}{\Delta_A} \end{matrix}$	$\begin{matrix} a & b \\ c & d \end{matrix}$

$$\Delta_z = z_{11}{\cdot}z_{22} - z_{12}{\cdot}z_{21} \qquad \Delta_h = h_{11}{\cdot}h_{22} - h_{12}{\cdot}h_{21} \qquad \Delta_A = A{\cdot}D - B{\cdot}C$$

$$\Delta_y = y_{11}{\cdot}y_{22} - y_{12}{\cdot}y_{21} \qquad \Delta_g = g_{11}{\cdot}g_{22} - g_{12}{\cdot}g_{21} \qquad \Delta_a = a{\cdot}d - b{\cdot}c$$

Since all of these parameter sets describe the same linear system, there is a one-to-one transformation between each parameter set. These transformations are shown in Table 7-1.

7.5 SCATTERING PARAMETERS [s_{ij}]

The scattering parameter set, commonly known as *s-parameters*, is different from the other two-port parameter sets. This must be true, since we have already exhausted all possible combinations of voltage and current relationships for two-port configurations. Instead of being a direct measurement of a voltage or current signal, the *s*-parameters measure the degree of separation of two signal components; each *s*-parameter is a ratio of an *incident* and a *reflected* signal, following from the definition of *scatter*, meaning "to go in different directions."

Scattering phenomena are easy to understand when scattered components are easily identified: for example, scattered light from atoms wherein incident light energizes the atom to re-emit scattered light. The intensity of the scattered light, relative to the incident light, is the scattering factor for the atom. Think of it as an *s*-parameter for that atom. The scattering factor for atmospheric gasses increases dramatically with frequency, such that the rate is almost ten times as great for blue light as for red light. This is why the sky appears to be blue while sunlight is basically white.

Scattering is more difficult to understand when such components do not exist. But, since we are dealing with linear systems, we can hypothesize any combination of signals (knowing that linear superposition works) and deal with these components separately. Of course, we would only go to this trouble if such definitions lead to useful relationships for designing and analyzing circuits. The definition that has proved useful can be illustrated with the case of a transistor driving a load: the "incident" component of the output signal is that which the transistor *would deliver* to a matched load (actually, a conjugate-matched load, since we are working with devices that are potentially reactive). Since the actual output signal will be different from this ideal situation, the difference between the actual value and the defined "incident" component is the definition of the "reflected" component. So, both input and output signals are the sum of their incident and reflected components.

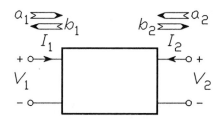

FIGURE 7-6 Schematic showing incident/reflected signals.

The generalized diagram in Figure 7-6 is similar to the previous two-port diagram, but augmented to show the incident (a_n) and reflected (b_n) components. Each port's voltage and current is the sum of its incident and reflected components

$$I_1 = I_{1i} + I_{1r} \qquad I_2 = I_{2i} + I_{2r}$$
$$V_1 = V_{1i} + V_{1r} \qquad V_2 = V_{2i} + V_{2r}$$

$(7-14)$

If we take the incident values as the independent values then the reflected values can be given by

$$b_1 = s_{11} \cdot a_1 + s_{12} \cdot a_2$$
$$b_2 = s_{21} \cdot a_1 + s_{22} \cdot a_2$$

$(7-15)$

or in a matrix form

$$\begin{bmatrix} b_1 \\ b_2 \end{bmatrix} = \begin{bmatrix} s_{11} & s_{12} \\ s_{21} & s_{22} \end{bmatrix} \begin{bmatrix} a_1 \\ a_2 \end{bmatrix}$$

$(7-16)$

The coefficients s_{ij} are dimensionless ratios of reflected values to incident values. For example, you could measure s_{11} by arranging for the output incident value to be zero. This would mean a_2 would be zero, and then the only influence on b_1 would be a_1. So, s_{11} is the ratio of b_1 to a_1. In fact, s_{ij} is the ratio of b_i to a_j where the incident value of the other port (a_i) is zero; or for all of the ratios

$$s_{11} = \left. \frac{b_1}{a_1} \right|_{a_2 = 0} = \begin{array}{l} input \\ reflection \\ ratio \end{array} \qquad s_{12} = \left. \frac{b_1}{a_2} \right|_{a_1 = 0} = \begin{array}{l} reverse \\ transmission \\ ratio \end{array}$$

$$s_{21} = \left. \frac{b_2}{a_1} \right|_{a_2 = 0} = \begin{array}{l} forward \\ transmission \\ ratio \end{array} \qquad s_{22} = \left. \frac{b_2}{a_2} \right|_{a_1 = 0} = \begin{array}{l} output \\ reflection \\ ratio \end{array}$$

$(7-17)$

Let's try measuring these parameters with PSpice.

The incident and reflected signals are related, by definition, to a "matched load" condition. This means that the scattering parameters are also related to whatever load, input and output, we consider being a matched load. So, before we can make these measurements, a load value must be chosen. For high-frequency work, 50 ohms is often chosen as a standard load impedance. We will use this value.

Since input and output loads are part of the problem, we will need to change the TEST subcircuit so that these loads are external:

```
.subckt TEST port1 port2

   xd1 port1 port2 3 DUT

   ibias   3 0 DC 5m
   cblck   3 0 1    ; one Farad
.ends
```

The input and output load will appear in our measurement circuits, attached to the TEST subcircuit and two voltage sources: an input source for the input load, and a power supply for the output load. Only the input source has an AC component, for measuring s_{11} and s_{21}, since the definition is that there be no incident signal at the output.

```
* measure s11 and s21

vli   1   0 DC 0 AC 1
rli   1   2 50
x1    2   3 TEST
rlo   3   4 50
vlo   4   0 DC 6 AC 0

ell   5   0 (2,0) 2 ; 2x input voltage
vll   11  5 AC -1   ; minus AC unity
rll   11  0 1G

e21   21  0 (3,0) 2 ; 2x output voltage
r21   21  0 1G
```

The measurement technique requires some explanation, and is simplified by the choice of a load value that has no reactance, since the conjugate value will be the same (by definition). Then we can manipulate the network theory formula for s_{11} and derive a useful result:

$$s_{11} = \left. \frac{b_1}{a_1} \right|_{a_2=0} = \frac{Z_{11} - Z_s^*}{Z_{11} + Z_s} = 2\left(\frac{Z_{11}}{Z_{11} + Z_s} \right) - 1 \tag{7-18}$$

where Z_s is the source load of 50 ohms, Z^* means the conjugate of Z (and equals Z, since the load impedance has no reactive component), and Z_{11} is the input impedance of the device. In the last section of (7-18), note that the ratio in the parentheses is like a voltage divider ratio: that is, the ratio is equal to the voltage at the input when the source voltage is unity. So we set the source voltage to unity and multiply the voltage measured at the input by two, then subtract one, to obtain the value of s_{11}. Similarly, we find that s_{21} can be obtained by multiplying the voltage measured at the output by two, since the input load matches the output load.

For measuring s_{12} and s_{22}, we effectively switch output for input. Now the output source has an AC component while the input has none, since the definition is that there be no incident signal at the input:

```
* measure s12 and s22

v2i   6   0 DC 0 AC 0
r2i   6   7 50
x2    7   8 TEST
r2o   8   9 50
v2o   9   0 DC 6 AC 1

e12  12   0 (7,0) 2 ; 2x input voltage
r12  12   0 50

e22  10   0 (8,0) 2 ; 2x load voltage
v22  22  10 AC -1   ; minus AC unity
r22  22   0 1G
```

The measurement technique is the same as before. As the calculations for the
s-parameters have been done with circuitry, we can simply print the values for each
parameter:

```
.print ac vm(11) vp(11) vm(12) vp(12)
.print ac vm(21) vp(21) vm(22) vp(22)
```

to provide a table of values versus frequency for s_{11} and s_{12}:

FREQ	VM(11)	VP(11)	VM(12)	VP(12)
1.000E+08	5.594E-01	-9.845E+01	1.889E-02	4.497E+01
2.000E+08	4.648E-01	-1.386E+02	2.251E-02	4.014E+01
3.000E+08	4.413E-01	-1.595E+02	2.468E-02	4.378E+01
4.000E+08	4.371E-01	-1.733E+02	2.693E-02	4.924E+01
5.000E+08	4.405E-01	1.762E+02	2.943E-02	5.491E+01
6.000E+08	4.481E-01	1.677E+02	3.219E-02	6.035E+01
7.000E+08	4.585E-01	1.603E+02	3.519E-02	6.548E+01
8.000E+08	4.711E-01	1.537E+02	3.844E-02	7.032E+01
9.000E+08	4.854E-01	1.478E+02	4.196E-02	7.492E+01
1.000E+09	5.013E-01	1.423E+02	4.580E-02	7.932E+01

and a table of values versus frequency for s_{21} and s_{22}:

FREQ	VM(21)	VP(21)	VM(22)	VP(22)
1.000E+08	8.902E+00	1.184E+02	8.293E-01	-9.224E+00
2.000E+08	5.096E+00	9.762E+01	7.718E-01	-7.761E+00
3.000E+08	3.510E+00	8.623E+01	7.576E-01	-7.158E+00
4.000E+08	2.674E+00	7.775E+01	7.536E-01	-7.229E+00
5.000E+08	2.163E+00	7.053E+01	7.534E-01	-7.657E+00
6.000E+08	1.819E+00	6.400E+01	7.550E-01	-8.294E+00
7.000E+08	1.573E+00	5.790E+01	7.580E-01	-9.069E+00
8.000E+08	1.387E+00	5.211E+01	7.620E-01	-9.947E+00
9.000E+08	1.242E+00	4.654E+01	7.668E-01	-1.091E+01
1.000E+09	1.124E+00	4.116E+01	7.725E-01	-1.195E+01

One caution: this technique for measuring the *s*-parameters relies on the input and
output loads being non-reactive and equal. Fortunately, these conditions are satisfied
by the common use of these parameters.

CHAPTER 8

LOOP GAIN ANALYSIS

Determining the loop gain of circuits with high-gain components is a difficult task. You might be tempted to open the loop to make measurements, but this will probably destroy the DC bias of the circuit. Opening the loop might also disconnect an internal load impedance and influence your measurements. What we need is a way to make these measurements **without opening the loop**[†] or changing the internal loading of the circuit.

In developing this technique we will start with a hypothetical circuit. We will set up this circuit to make these measurements without introducing any problems. Then we will modify the formulas to work with "real" circuits. But first, let us review some terms.

8.1 AN IDEAL CIRCUIT

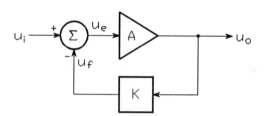

FIGURE 8-1 Schematic of system with feedback.

In a system with feedback, shown in Figure 8-1, the following signals may be calculated by inspection:

$$u_o = A \cdot u_e$$
$$u_e = u_i - K \cdot u_o$$

$$(8\text{-}1)$$

where u_i, u_o, and u_e are the *input*, *output*, and *error* signals, respectively. Further

---

[†] This technique, as far as the author can determine, originates from a *Hewlett-Packard Applications Note*, circa 1965. The technique has been extended and refined by Dr. R. D. Middlebrook of the California Institute of Technology; see *International Journal of Electronics*, vol. 38, no. 4 (1975), 485-512.

manipulation of these formulas yields

$$u_o = A \cdot u_i - A \cdot K \cdot u_o$$

$$u_e = \frac{u_i}{1 + A \cdot K} \qquad (8\text{-}2)$$

so that system gain, G, is

$$G = \frac{u_o}{u_i} = \frac{A}{1 + A \cdot K} = \frac{A}{1 + T} = \frac{1}{K}\frac{T}{1 + T} \qquad (8\text{-}3)$$

where the *loop gain*, or *return ratio*, T is

$$T = \frac{u_f}{u_e} = A \cdot K \qquad (8\text{-}4)$$

The system's relative stability can be inspected in an open-loop configuration. The loop is opened in the feedback path, and a "test" signal is injected. Then the resulting feedback signal, opposite the injection point, is compared to the test signal. The feedback signal is inspected for one full cycle (360°) of phase shift. However, since the feedback is subtracted from the input, the subtraction alone provides 180° of phase shift. So the feedback signal should be inspected for an additional 180° of phase shift, and not 360°. If the loop gain (the ratio of feedback signal to the test signal) is one, or greater, the loop is unstable, since it can supply its input. The amount of gain, relative to unity, at 180° phase shift is called the loop *gain margin*. Likewise, the amount of phase difference from 180°, when the loop gain is unity, is called the loop *phase margin*.

This analysis is true even if the loop is broken in the forward path, but it seems easier to describe as though the feedback path was broken. Now, having said that, we will imagine that we will "break" the loop somewhere inside the circuit. It doesn't matter where, just as long as the break is in some part of the signal path. To do this, we will imagine that some part of the signal path is a controlled-current source connected to an impedance (for example, imagine an ideal transistor with a load resistance) as shown in Figure 8-2. Also, the normal input signal will be set to zero so that we need only consider the effects of the test signal.

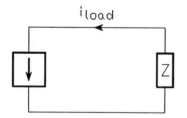
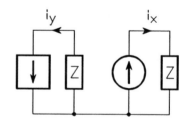

FIGURE 8-2   Idealized section of loop circuitry.   FIGURE 8-3   Load duplicated, current injected.

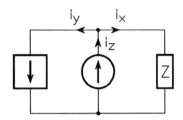

FIGURE 8-4 Via superposition, loop is not broken.

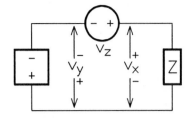

FIGURE 8-5 Voltage-mode equivalent.

Then we break the loop, as shown in Figure 8-3, being careful to duplicate the load impedance seen by the current source. Now we inject a current into the original load impedance, which is part of the original loop, as the test signal. The return signal is current in the duplicated load impedance. The ratio of the two signals is merely the **current loop gain** measurement, or Ti, where $Ti = i_y/i_x$.

Now comes part of the trick: it is not necessary to open the loop to inject the test signal. As shown in Figure 8-4, if a current is injected directly into the signal path, it splits into the test signal and return signal of Figure 8-3. Furthermore, the load impedance does not need to be duplicated, since the loop is not broken and the current source has infinite impedance.

Similarly, we can develop the same technique using voltages instead of currents. As shown in Figure 8-5, a voltage source is inserted in the loop to inject a signal. The resulting measurement of the voltage across the load impedance and the controlled-voltage source yields a ratio that is merely the **voltage loop gain** measurement, or Tv, where $Tv = v_y/v_x$.

Since we were able to choose ideal points to break the loop, the signal ratios are the loop gain of the system; that is, $T \equiv Ti$ for the current measurement, and $T \equiv Tv$ for the voltage measurement.

8.2 A "REAL" CIRCUIT

Now we tackle a "real" circuit. Our example so far assumed controlled sources that have no internal impedance the way a real transistor, or real opamp, does. We can account for this impedance by use of superposition.

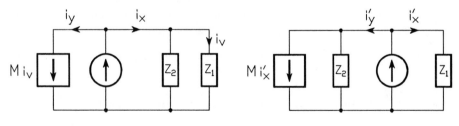

FIGURE 8-6 Ideal injection.

FIGURE 8-7 Real injection.

In the previous case the ideal injection was performed as shown in Figure 8-6, but in this case we have two new currents, as shown in Figure 8-7:

$$i'_x = \frac{Z_2}{Z_1+Z_2} i_x = \frac{i_x}{1+\dfrac{Z_1}{Z_2}}$$

$$i'_y = i_y + \frac{Z_1}{Z_1+Z_2} i_x = M \cdot i'_x + \frac{Z_1}{Z_2} i'_x \tag{8-5}$$

where M is the rest of the loop's gain. The current loop ratio that we measure, then, is

$$Ti = \frac{i'_y}{i'_x} = M + \frac{Z_1}{Z_2} \tag{8-6}$$

and the loop gain is

$$T = \frac{i_y}{i_x} = \frac{M \cdot i'_x}{\left(1+\dfrac{Z_1}{Z_2}\right) i'_x} = \frac{M+\dfrac{Z_1}{Z_2}-\dfrac{Z_1}{Z_2}}{1+\dfrac{Z_1}{Z_2}} = \frac{Ti-\dfrac{Z_1}{Z_2}}{1+\dfrac{Z_1}{Z_2}} \tag{8-7}$$

Notice that this is the measurement of a real circuit, where the active element has been replaced with its Norton-equivalent current source and impedance, Z_2. The remaining impedance, Z_1, represents the load of the active element.

By similar means, we might make a non-ideal measurement of the voltage loop ratio, as shown in Figure 8-8, with the result

$$T = \frac{Tv-\dfrac{Z_2}{Z_1}}{1+\dfrac{Z_2}{Z_1}} \tag{8-8}$$

Again, notice that this is the measurement of a real circuit, where the active element has been replaced with its Thévinin-equivalent voltage source and impedance, Z_2. The remaining impedance, Z_1, represents the load of the active element.

Of course, this begs the question of what these circuit impedances are so we may calculate T exactly. By inspection, we can see that

$$T \approx Ti, \quad \text{if } \frac{Z_1}{Z_2} \ll 1 \text{ and } \frac{Z_1}{Z_2} \ll T$$

$$T \approx Tv, \quad \text{if } \frac{Z_2}{Z_1} \ll 1 \text{ and } \frac{Z_2}{Z_1} \ll T \tag{8-9}$$

Notice that if we were to measure both Ti and Tv, we would have two equations with two unknowns. To eliminate the impedance ratio, we first rewrite the measurement

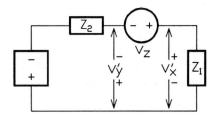

FIGURE 8-8 Real injection for voltage measurement.

ratios in terms of T

$$Ti = \left(1 + \frac{Z_1}{Z_2}\right)T + \frac{Z_1}{Z_2}$$

$$Tv = \left(1 + \frac{Z_2}{Z_1}\right)T + \frac{Z_2}{Z_1}$$

(8-10)

and then, by adding 1 to both sides and adding the reciprocals, we find that

$$(T+1) = (Ti+1)\,||\,(Tv+1)$$

(8-11)

where $||$ means "parallel combination"; for example:

$$x\,||\,y = \frac{1}{\dfrac{1}{x} + \dfrac{1}{y}} = \frac{x\,y}{x+y}$$

(8-12)

This says that the lower of the two measurements, Ti or Tv, dominates the value of T, the loop gain. It also says that we can make both measurements, as shown in Figure 8-9, and calculate T exactly. Another way to restate the formula for T is

$$T = \frac{Ti \cdot Tv - 1}{Ti + Tv + 2}$$

(8-13)

which may be more suitable for numerically stable calculations.

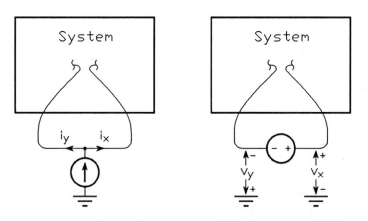

FIGURE 8-9 Making Ti and Tv measurements.

8.3 A "REAL" EXAMPLE

Let's try to calculate loop gain in a relatively simple circuit. Figure 8-10 shows the circuit we will measure, breaking into the loop at the output of the opamp.

The simplified model of the opamp (also shown) will be used for this example. The subcircuit definition for the opamp is

```
* "ideal" op-amp with 100K gain and one-pole roll-off at 10Hz
.subckt opamp non inv out
  rin non inv 100K
  egain 1 0 (non,inv) 100K
  ropen 1 2 1K
  copen 2 0 15.92u
  eout  3 0 (2,0) 1
  rout  3 out 50
.ends
```

Since we will want two copies of the entire circuit we are measuring, let's put the circuit in a subcircuit. This subcircuit will have only two nodes, which are at the place where we are breaking the loop:

```
* example circuit
.subckt test left right
  vin 1 0 DC 0
  x1 1 2 left opamp
  r1 right 0 200
  r2 right 2 10K
  r3 2 0 1K
  c1 2 0 .038u
.ends
```

Finally, there is the rest of the circuit:

```
* Loop gain measurement
.ac dec 100 1 1Meg
.probe
xi Ti_left Ti_right test ; this copy for Ti measurement
xv Tv_left Tv_right test ; this copy for Tv measurement

* perform Ti measurements
iz 0 1 AC 1 ; current stimulus
viy 1 Ti_left  DC 0 ; sense Ix
hiy iy 0 viy 1       ; convert Ix to a voltage
riy iy 0 1G
vix 1 Ti_right DC 0 ; sense Iy
hix ix 0 vix 1       ; convert Iy to a voltage
rix ix 0 1G

* perform Tv measurements
vz Tv_right Tv_left AC 1 ; voltage stimulus
evy vy 0 (0,Tv_left ) 1 ; duplicate Vx
rvy vy 0 1G
evx vx 0 (Tv_right,0) 1 ; duplicate Vy
rvx vx 0 1G
```

Notice that we may break into the loop of this example circuit elsewhere, or try another circuit, just by changing the description of the test subcircuit.

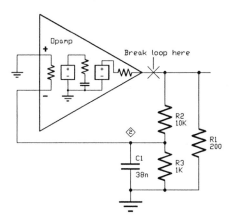

FIGURE 8-10 Loop gain measurement example circuit.

After running an AC analysis simulation, we view the results with the aid of the following macro definitions. (Probe macro definition and use are covered in §6.6.)

```
par(a,b)=(((a)*(b))/((a)+(b)))
Ti=(V(iy)/V(ix))
Tv=(V(vy)/V(vx))
T=(par(Ti+1,Tv+1)-1)
```

The first macro, par(a,b), defines the "parallel" operation (that is, $a \,||\, b$) and is written to be numerically stable as the arguments approach zero. Without macros, or Probe handling complex arithmetic, displaying the loop functions becomes quite an ordeal. For example, the relatively simple expression for the magnitude of Ti+1 is

$$\left| \text{Ti+1} \right| = \sqrt{\frac{\left(\text{i}x_{re} - \text{i}y_{re} \right)^2 + \left(\text{i}x_{im} - \text{i}y_{im} \right)^2}{\text{i}x_{re}^2 + \text{i}y_{im}^2}} \qquad (8\text{-}14)$$

where, for example, $\text{i}x_{re}$ is the real part of $\text{i}x$.

First we look at the magnitude response of T, Ti, and Tv, as shown in Figure 8-11. As we would expect, at lower frequencies Tv dominates, since the input impedance of the feedback circuit is much greater than the output impedance of the opamp; that is, Tv is of smaller magnitude and will control the value of $(\text{Ti}+1) \,||\, (\text{Tv}+1)$. But as the frequency increases, the impedance of the feedback circuit decreases and begins to load the opamp. As the loop gain components, Ti and Tv, approach unity, or 0dB, their contribution is dominated by the +1 in the calculation of T derived in (8-11). Weird and non-intuitive things happen to Ti and Tv as frequency increases, but T has the shape we expect from knowing the frequency response of the opamp and the feedback circuit.

Now we look at the phase response of T, Ti, and Tv, as shown in Figure 8-12. Here we find that, at low frequencies, the phase is controlled by the opamp. But again, as frequency increases, the current and voltage components, Ti and Tv, give

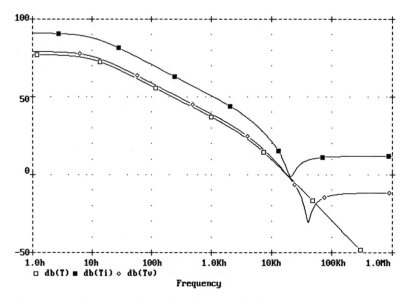

Figure 8-11   Plot of open-loop magnitude responses.

an inaccurate indication of the loop phase response. Intuitively, we know that shape of T is correct from knowing the opamp's phase response, which has a single pole, and that the feedback circuit has no resonant circuit that would give the 180° phase shift of a double pole.

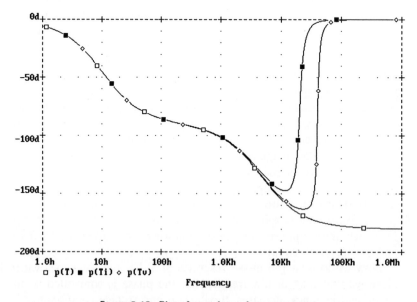

Figure 8-12   Plot of open-loop phase responses.

8.4 UNSTABLE LOOP GAIN

Every once in a while, through no fault of the engineer (of course), a system is designed that is unstable or only conditionally stable. The problem then becomes how to correct this, but the type and amount of correction needed depend on how "bad" the system is. How do we ascertain this?

The key to measuring an unstable loop comes from reconsidering the Ti and Tv measurements we made earlier. As shown in Figure 8-7 and Figure 8-8, in each case two signals are created and measured. The ratio of the signals is some type of loop gain measurement. Analogues of those figures are shown in Figure 8-13 and Figure 8-14, where now we assume the generating sources injecting the test signal have finite impedance. However, we can easily see that this does not affect the calculation of Ti or Tv. In both cases, the ratio of the current or voltage values, now *primed* values in the newer figures, will be the same as before.

But the circuit has changed! The impedance of the test signal source changes the loading on the active device, yet the loop gain measurement is not changed. This happens because the effect of the source impedance is not part of the measured values. Another way of saying this is that the measurements have been designed to be taken outside the subnetwork that contains the source impedance.

But the source impedance still loads the circuit! This gives us a mechanism for altering the circuit's response without affecting the original loop gain measurement. In particular, we can use the source impedance, or an additional impedance associated with the source, to lower the loop gain so that it becomes marginally stable. Then we make our measurements to calculate the loop gain.

In practice, forcing the loop to be stable is only a concern for real circuits. Before starting small-signal analysis, the simulator calculates an operating point ignoring all capacitances and inductances, which are the effects that usually create instability. PSpice allows you to simulate and measure an unstable loop gain without resorting to a loading impedance to stabilize the loop, usually. However, you might come across a circuit that is DC unstable, having net positive feedback at zero frequency. In this case, PSpice will have problems finding an operating point and you need to use the technique just discussed to analyze such a circuit.

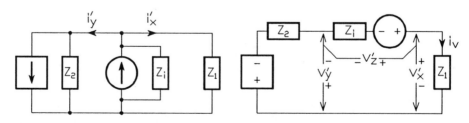

FIGURE 8-13 Current-mode measurement.   FIGURE 8-14 Voltage-mode measurement.

The first applications of feedback control theory were in electrical systems, in contrast to earlier mechanical systems that were often designed using Routh's method as the only analytical tool. During the early days of electrical and electronic systems a great amount of effort was spent on analytical techniques for these systems. The result was a variety of graphical tools for analyzing the frequency response of systems and synthesizing stable control systems, both electrical and mechanical. The remarkable outcome of these tools is the quick, graphic analysis of transient, or time domain, response and stability from frequency response. Moreover, that closed-loop frequency response can be deduced from open-loop frequency response.

Most physical systems, or *plants* (a control theory idiom), may be analyzed by PSpice by creating an electrical analog for the controlling equations: force translates into current, mass translates into capacitance, and so forth. Then the frequency response of the system (linear displacement, rotation, etc.) may be calculated using PSpice's frequency response analysis. As mentioned in the introduction to this book, PSpice is not concerned that your electrical analogue is not even a normal electronic circuit. By using this technique of "analysis by analogy" you may directly convert system equations in "*s*" to electrical equivalents for PSpice.

9.1 DYNAMIC PLANT EXAMPLE

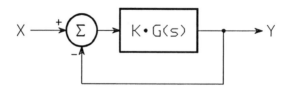

FIGURE 9-1   Schematic of a generalized feedback system.

Let us consider a general example of a linear system with feedback control, as shown in Figure 9-1. This system has a "dynamic plant" and simple feedback. It is comprised of

- an input signal, X,
- an error signal, E,
- a "plant" function, K·G(*s*), which separates the linear gain from the complex part of the plant, and
- an output signal, Y.

Of course, when the loop is open the error signal is the same as the input signal. Functions of these signals are calculated to create the various plots in this chapter; for example:

$$\frac{|Y|}{|X|} = \text{system magnitude response}$$

$$\angle Y - \angle X = \text{system phase response}$$

Let us look at an example circuit and see what feedback system analysis plots result from its frequency response.

Figure 9-2 shows a more complicated example of the two-pole filter we analyzed in Chapter 6. The circuit file that describes this circuit is

```
2 tandem, double-pole, low-pass, LC-filters
Vin 1 0 AC 1
Rin 1 0 1K
Ein in 0 poly(2) (1,0) (6,0) 0 2 0
*   Q = 4
R1 in 2 25
L1 2 3 10mH
C1 3 0 1uF
*   Q = 4 @ 1/5 frequency
E2 4 0 3 0 1
R2 4 5 25
L2 5 6 50mH
C2 6 0 5uF
.AC DEC 100 100hz 10Khz
.PROBE
.END
```

You will notice that we have placed in series two of the "Q = 4" RLC filter sections we analyzed earlier. We have also used an E device to serve as an ideal difference amplifier, which lets us change the gain easily and choose to close the feedback loop. In particular, if the gain from the feedback input is zero, the circuit is operating open-loop. This circuit does not correspond to any useful circuit or physical system analogy. However, we can still look at the system response.

9.2 BODE PLOTS

An important theorem by H. W. Bode deals with linear systems having constant coefficients with poles and zeros in the left-half of the "*s*-plane" only (minimum-phase systems). This theorem holds that, for any minimum-phase system, the phase-angle part of the frequency response is uniquely related to the magnitude part. This allows the phase response of a circuit to be deduced from the magnitude response, and vice versa. Moreover, entire systems of coupled minimum-phase sections could be analyzed by a graphical technique that embodies Bode's theorem. This technique was extremely useful in analyzing electronic circuits, where complex systems can be easily built but are potentially more difficult to analyze. The graphical technique used what came to be called *Bode plots*.

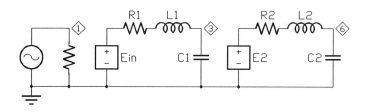

FIGURE 9-2 Example circuit: tandem LC circuit with gain.

Bode plots are graphs of the magnitude and phase response of the circuit versus the frequency of sinusoidal excitation (that is, small-signal analysis, or what SPICE calls frequency response). We have seen these plots in the previous chapter. The simplicity of construction for these plots, **when done by hand**, is due to the frequency scale (X-axis) being logarithmic. Phase angle is plotted linearly along the Y-axis, but magnitude is plotted with a logarithmic scale, usually in decibels (dB), along the Y-axis. This setup allows ratios of magnitude, ratios of frequency, and differences in phase to have constant displacement on the graph.

The ease in constructing Bode plots made them popular and standard. PSpice creates them for you either by using the .PLOT statement or, more usefully, by using Probe graphics. Now, after running PSpice on this circuit we may run Probe to look at some Bode plots.

The plot in Figure 9-3 shows the open-loop magnitude response of the system. We can see that system response is the product of its components: a gain section, a lower-frequency two-pole filter, and a higher-frequency two-pole filter. We may plot these responses separately, as shown in Figure 9-4.

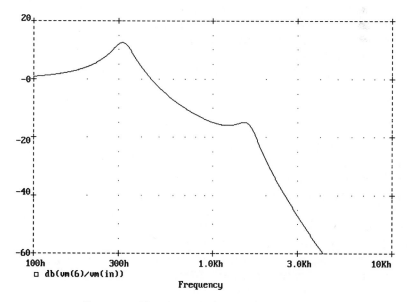

FIGURE 9-3 Plot of open-loop magnitude response.

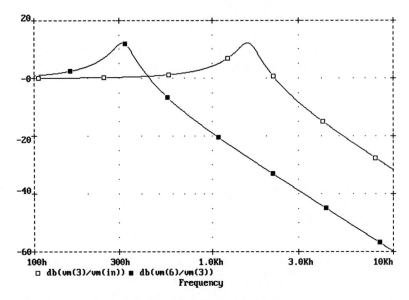

FIGURE 9-4 Plot of individual section's response.

Phase plots are also simple, as shown in Figure 9-5. Notice that the total phase is the sum of the phase of the individual sections, also shown in Figure 9-5.

These examples show what made the Bode plot technique so useful and popular: the ease with which graphs of the response of each section's circuitry could be

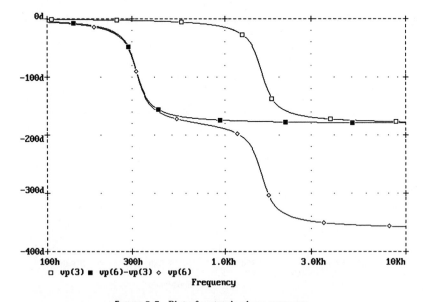

FIGURE 9-5 Plot of system's phase response.

constructed and then combined to produce the system response graph. Of course, this was all being done **by hand**, but now you can use PSpice.

One of the items not handled well by Bode plots was deducing closed-loop response from open-loop response. We can re-simulate our circuit in the closed-loop configuration by modifying the statement for the difference amplifier (E device) as follows:

```
Ein in 0 poly(2) (1,0) (6,0) 0 2 0
```

becomes

```
Ein in 0 poly(2) (1,0) (6,0) 0 2 -2
```

to close the loop. After running PSpice we may compare the open-loop and closed-loop responses, as shown in Figure 9-6. You will notice that the system gain peaks at a new frequency and is the frequency at which the open-loop response crossed 0dB!

As you might expect, the phase response of the system has changed as well. This is shown in Figure 9-7.

In the next section we will look at another type of plot that does deduce closed-loop response, from the open-loop response, by merely shifting the origin of the plot.

9.3 INVERSE-POLAR (NYQUIST) PLOTS

Another graphical technique, developed by H. Nyquist, uses a polar coordinate graph to plot system response characteristics. From the mathematics of feedback systems, we know that the closed-loop system response is

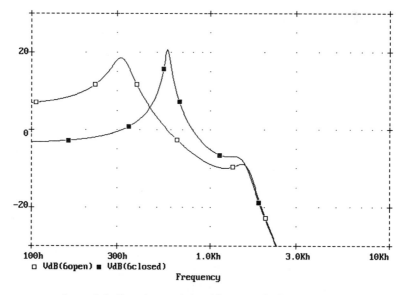

FIGURE 9-6 Open-loop and closed-loop magnitude response.

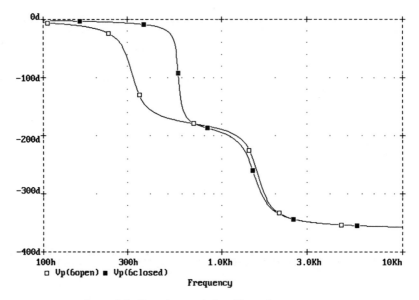

FIGURE 9-7 Open-loop and closed-loop phase response.

$$\frac{Y}{X} = \frac{K \cdot G(s)}{1 + K \cdot G(s)} \tag{9-1}$$

The magnitude and phase of this response can be plotted for various frequencies, as with the Bode plot, if we could plot this response using polar coordinates. In this case the magnitude of the response would be the magnitude of the vector, and the phase response would be the angle of the vector.

Such a plot is quite easy to create with Probe, since the magnitude and phase of a vector are functions of the real and imaginary parts of the vector. This is called a *polar plot* and is created directly using the R( ) and IMG( ) display functions, which we used previously (for instance, when calculating impedances). By changing the X-axis of the plot to be the real component of the vector and plotting the imaginary component, the polar plot of the vector is created by using the rectangular components of the vector. This is shown in Figure 9-8. The only parts missing from the plot are the markers for frequency.

In the feedback system equations, notice that

$$K \cdot G(s) = \frac{Y}{E} \tag{9-2}$$

and that the inverse of the system function is

$$\left(\frac{Y}{X}\right)^{-1} = \frac{X}{Y} = \frac{1}{K \cdot G(s)} + 1 = \frac{E}{Y} + 1 \tag{9-3}$$

which is very convenient, since E = X when the system is open-loop. This means that the closed-loop response is related to the open-loop response by adding unity, when

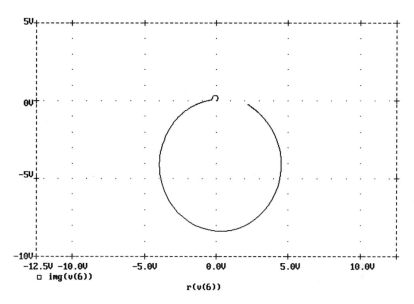

FIGURE 9-8 Open-loop response in polar form.

the response is plotted as an inverse function. This is the utility of the "inverse polar" plot: **the plot of the inverse transfer function is the same for both open-loop and closed-loop configurations, only the origin of the plot shifts.** Let us try this with our example circuit.

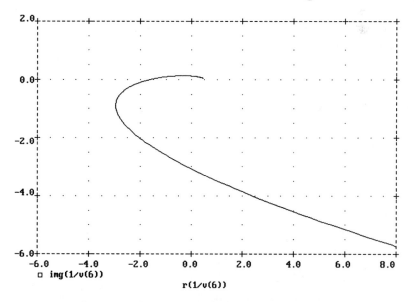

FIGURE 9-9 Open-loop response in inverse polar form.

Using the open-loop version of our circuit file (no feedback through `Ein`), the inverse polar form of the output response is shown in Figure 9-9. Then we re-simulate using the closed-loop version of the circuit file. Using the same display functions as before, the closed-loop response is plotted in Probe as shown in Figure 9-10.

If you are trying this, you will probably need to adjust the X-axis and Y-axis ranges, zooming in to the origin, to scale the plots for use and comparison. With everything adjusted, we notice that the curve is the same and only the origin of the plot has changed between the open-loop and closed-loop response plots.

EXERCISE 9.3-1

Check our equations and PSpice: directly evaluate E/Y + 1 in Probe and compare the open-loop response to the closed-loop response.

9.4 NICHOLS PLOTS

The Bode plot has the advantage of rapid plotting **by hand** using asymptote approximations and multiplying functions together by adding distances on a log scale. The inverse polar plot is more difficult to assemble by hand, but once plotted shows both the open-loop and closed-loop response. There is another graphical technique, developed by N. B. Nichols, which has both features.

The Nichols plot shows both magnitude and phase on the same plot, and is like a combined Bode plot. Usually, the easiest way to construct a Nichols plot by hand

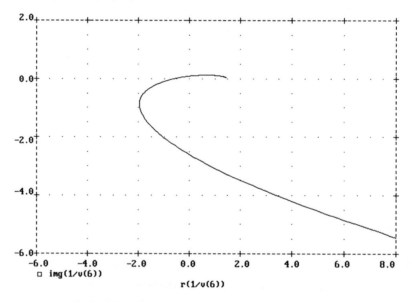

FIGURE 9-10 Closed-loop response in inverse polar form.

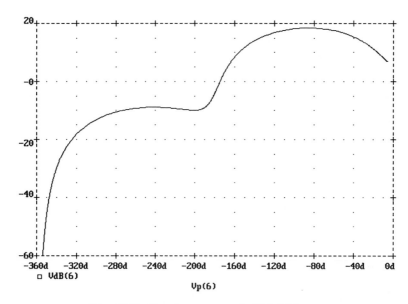

FIGURE 9-11 Open-loop response in Nichols' form.

is to first construct a Bode plot and transcribe points to the Nichols plot. Using our example, we obtain the Nichols plot shown in Figure 9-11 by resetting the X-axis variable to be the phase response of the system.

The Nichols plot technique then uses an overlay plot to transform the open-loop response into a closed-loop response. This overlay plot is the same (so long as the plot coordinates are the same) for all system responses and performs the same job as shifting the origin for the inverse-polar plot.

CHAPTER 10

NOISE ANALYSIS

Whenever small signals are amplified or measured, you usually reach a lower limit of signal that is discernible; this limit is set by spontaneous fluctuations in the equipment you are using. The spontaneous fluctuations are called *noise*, since, if the audio-frequency component of the fluctuating voltage, or current, were amplified and fed into a loudspeaker, you would hear a hissing noise. This type of fluctuation extends across all frequencies; for example, in television equipment noise creates the "snowy" picture.

PSpice can analyze noise by calculating the noise contributions from each element in the circuit and combines these noise sources with the various transfer functions in the circuit. You may print or plot (or use Probe) the result of these calculations to obtain noise response over a range of frequencies, just as you did with the frequency response analysis.

10.1 NOISE CALCULATIONS

In PSpice, the resistors and semiconductor devices contribute to the noise calculations. While the semiconductor device noise models are more complicated, we can understand the general idea of noise analysis just by using resistors. Later the semiconductor noise models will be explained.

Thermal noise is due to the random motion of a large number of electrons that cause an interaction between free electrons and vibrating ions in a conducting medium (such as a resistor). Similarly, random excitations of polarizable molecules, forming dipoles, cause thermal noise in a lossy dielectric (such as a capacitor). Thermal noise is also called *Johnson noise* due to thorough experimental studies made by J. B. Johnson.

The resistor, in PSpice, generates an equivalent thermal noise current in parallel with the resistor (which is then noiseless). Thermal noise is *white*: its fluctuations contain equal amounts of all frequency components, which, in human vision, is the sensation of white light. Technically, we would say thermal noise has "constant spectral density." The random fluctuations are characterized on a statistical basis using averages so that, while the time average of a random fluctuation is zero, its mean-square average deviation (or *variance*) has a value. The level of the resistor's current generator is

$$\overline{i^2} = \frac{4 \cdot k \cdot T}{R} \Delta f \tag{10-1}$$

where[†]

k = Boltzmann's constant $\approx 1.38E{-}23$ (W·sec/°K)

T = temperature (°K)

R = resistance (Ohm)

Δf = bandwidth (Hertz)

Alternatively, the equivalent thermal noise could be represented by a mean-square voltage source in series with the resistor (which would then be noiseless) with the level

$$\overline{e^2} = \overline{i^2}{\cdot}R^2 = 4{\cdot}k{\cdot}T{\cdot}R{\cdot}\Delta f \qquad (10\text{-}2)$$

The two methods are equivalent; however, the "series voltage" technique adds another node to the circuit. Thus, the "parallel current" technique is used.

For the entire circuit, each noise generator's contribution is calculated and propagated, by the appropriate transfer function, to the output of the circuit. At the output, all of these contributions are RMS-summed to obtain the "total output noise." RMS stands for Root Mean Square: each contribution is squared, then the square root is taken of the average (mean) of all these contributions. This is the formula for adding variances. Also, since the transfer function from output to anywhere else in the circuit is known, PSpice will calculate the "equivalent input noise." Detailed reports may, optionally, be generated showing the individual noise contributions.

10.2 THE .NOISE STATEMENT

The .NOISE statement directs PSpice to perform the noise calculations and specifies which nodes are the output, and where the input is. Noise calculations are done over a range of frequencies and **are done in conjunction with a frequency response analysis**. This means that both .NOISE and .AC statements are required to do the noise calculations. The .AC statement sets the frequencies at which the noise calculations are done. The statement that controls the noise calculations is

.NOISE V(*node* [[,*node*]]) *source_name* [[*interval*]]

where V(*node* [[,*node*]]) is the total output noise voltage; it may be a single node, in which case the noise voltage is referenced to ground, or a pair of nodes, in which case the noise voltage is taken to be across the two nodes. *source_name* is the name of an independent source (V device or I device) to which the total output noise will

---

[†] In the noise formulas, for $f \gg 1\text{GHz}$, quantum effects require the quantity $k\text{T}$ be replaced by: $\dfrac{hf}{e^{\frac{hf}{k\text{T}}}-1}$

be referred when calculating equivalent input noise. This source is not a noise generator; it is just a reference for describing the input (most likely it is the input for your frequency analysis).

If present, *interval* causes the detailed printout of individual device noise contributions. As you may have guessed, this happens every *n*th frequency where *n* is the interval value. The individual contributions are referred to the output, so you may judge how important each is to the overall noise performance of the circuit, and are not the noise amounts for each contributor. The detailed printout, if specified, is generated regardless of any other output you might designate.

10.3 PRINT AND PLOT OUTPUT

Output from noise analysis may be generated by `.PRINT` or `.PLOT` statements, just as in AC analysis. In either case, the output is organized by the frequency at which the calculations were made. The statement forms are

> `.PRINT NOISE` *output...*

and

> `.PLOT NOISE` *output...*

Each *output* entry becomes a column in the table output by the `.PRINT` statement, or a curve in the plot output by the `.PLOT` statement. The output values you can print/plot are predefined as

`ONOISE`	total noise at the designated output
`INOISE`	`ONOISE` referred to the input source
`DB(ONOISE)`	`ONOISE` in decibels (referred to 1-volt/Hertz$^{1/2}$)
`DB(INOISE)`	`INOISE` in decibels (referred to 1-volt/Hertz$^{1/2}$)

10.4 GRAPHICS OUTPUT

Probe can also graph the values `INOISE` and `ONOISE`, showing these values in decibels, which makes it easy to display results and make calculations. Let us take a look at a simple example of noise as demonstrated in a phonograph system to highlight the limits noise imposes.

Most of the noise of an amplifying system is introduced in the first stage of the system, and, of course, once the noise is introduced nothing can be done to reduce its effects. In a phono-preamplifier, even if the amplification of the signal from a magnetic pickup were noise-free, there is still thermal noise generated by the magnetic cartridge, itself, and by the resistive load for which it was designed. We may model the phono-cartridge and preamplifier as shown in Figure 10-1.

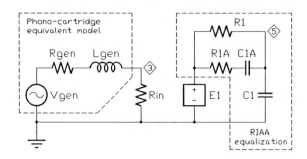

Figure 10-1 Schematic of phono amplifier circuit.

The amplifier, E1, is assumed to be noiseless. The section marked "RIAA equalization" refers to a standard pre-emphasis that was involved in cutting the master record mold. This pre-emphasis was designed to limit the excursion of the groove at low frequencies and to boost signal strength at high frequencies. The preamplifier has to undo this pre-emphasis, with an opposite de-emphasis, and in doing so modifies the noise results from what you would have for an amplifier without de-emphasis. The circuit file for this follows:

```
Noise from magnetic phono-cartridge
Vgen 1 0 AC 1
Rgen 1 2 1350
Lgen 2 3 .5
Rin  3 0 47K
E1   4 0 3 0 10 ; first pole of RIAA curve @ 50Hz w/20dB boost
R1   4 5 1
C1   5 0 3.528m
R1A  4 6 212.8m ; pole @ 2120Hz
C1A  6 5 352.8u ; zero @ 500Hz
.ac dec 100 20 20K
.noise v(5) Vgen 100
.probe
.end
```

You will notice that we have selected the output of the equalization as our noise output and the signal generator of the phono-cartridge as our reference input. We have selected a frequency response analysis with a range of 20 Hertz to 20,000 Hertz to cover the entire audio range. This is the range over which the noise calculations will be made. After running PSpice, we may graph the noise results directly. But first, let's look at the transfer function of the equalizer, so you will understand some of the noise calculations we try later. The transfer function is displayed in Figure 10-2.

With 1,000 Hertz as the unity gain reference frequency, you can see that the lower frequencies get quite a boost. Let us see how this changes the noise from our pre-amplifier. As shown in Figure 10-3, the boost in the equalization circuit amplifies the noise from the phono-cartridge and input load. Fortunately, the human ear is not as distracted by noise at lower frequencies as those in the mid-range (where the noise has already approached its lowest value).

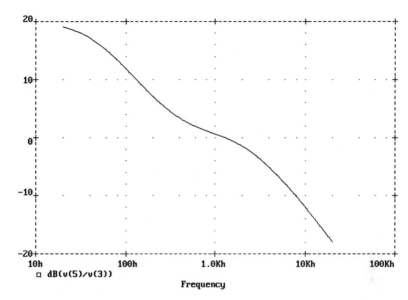

FIGURE 10-2 Plot of RIAA equalization curve.

10.5 CALCULATING TOTAL NOISE AND S/N

Using the phono-cartridge example from the previous section, we will use Probe to directly calculate noise totals. Total noise is the overall variance of the combined

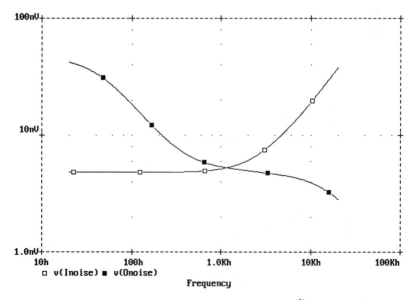

FIGURE 10-3 Noise density plots (volt/Hertz$^{1/2}$).

noise fluctuations at each frequency. This is the RMS calculation we discussed earlier. In Probe, so long as the frequency range for the calculation has been simulated, we can directly calculate total noise as shown.

The plot in Figure 10-4 is the running total of the noise contributions at each frequency, so that the right-most point on the graph is the total for the entire range of frequencies. The cursors are used, in Probe, to measure the total RMS noise across the band. The total noise is the y-value of cursor "C1," in this case 646.2nV. Although the greatest noise values occurred at low frequencies, we can see that the higher frequencies made up for this by having more bandwidth! Even so, nearly one-half of the total noise comes from the 20-to-200 Hertz band.

Now that we can calculate total noise, calculating *signal-to-noise* (S/N) is similar, since

$$S/N \; = \; 20 \cdot log \left(\frac{signal}{total\ noise} \right) \tag{10-3}$$

That is, the value for S/N is the ratio of signal power to noise power expressed in decibels. Given that our magnetic phono-cartridge, in this example, generated an average signal (after de-emphasis) of 4mV, we may now display the graph for S/N. If you do this often, you might consider creating a Probe macro for S/N; for example:

```
SN(signal) = ((signal)/sqrt(s(v(onoise)*v(onoise))))
```

This was done for the plot in Figure 10-5, which shows the running result for S/N, starting at 20 Hertz. We can see that for the entire audio range, the limiting signal-to-noise ratio to expect from an ideal pre-amplifier is under 80dB.

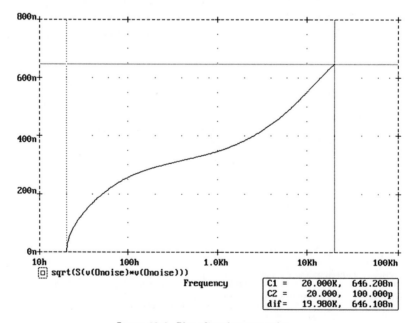

Figure 10-4 Plot of total output noise.

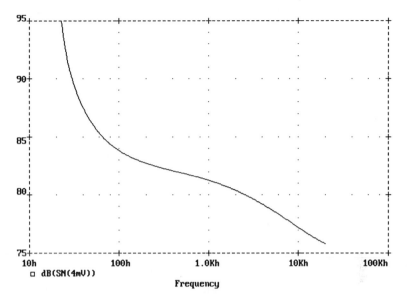

FIGURE 10-5 Plot of signal-to-noise (S/N).

10.6 INSERTING NOISE SOURCES

So far, we have considered only the noise generated by resistors. The active devices also have noise associated with their operation, which we will review in Chapter 16 on the semiconductor devices. But what if you want to insert a noise source directly? For example, you may want to enhance the modeling of an idealized circuit by providing for noise. This may be accomplished by using the controlled sources to "insert" noise.

The noise analysis done by PSpice will calculate the value of the noise voltages, or currents, referenced to the input you have selected. This is a way of modeling noise for the entire circuit where the circuit is assumed to be noiseless, with all of the noise being generated by a noise source at the input to the circuit. Using this approach makes it easy to compare the merits of one circuit over another when comparing noise specifications. You will see this in the data books, say, for operational amplifiers where the amplifier has a specified, nominal gain, as well as a specified input noise voltage and current. Of course, the values for the noise sources vary with frequency, but for most work the constant values may be assumed. The model for an operational amplifier, with noise sources included, is shown in Figure 10-6.

The noise voltage is a voltage source in series with the input, since the operational amplifier is a voltage amplifier (that is, it has a high input impedance). The noise current is in parallel with each input and the effect of these noise sources is determined by the resistance of the input circuit (such as a transducer). In accordance with Ohm's law, the value of the noise current is converted to a noise voltage by the value of the resistance, or impedance. If the input to the amplifier has

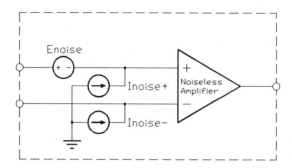

FIGURE 10-6 Schematic of opamp noise model.

a low impedance, the noise generated by the current noise sources will add little to the overall input noise voltage. This would be the case if, say, the input to the amplifier were the output of another amplifier that has a fairly low output impedance. However, if the circuit generating the input signal to the amplifier were of high impedance, such as a crystal microphone, the input current noise will be much greater. The limiting value of the noise contributed by the noise currents is determined by the input impedance of the amplifier.

Creating models for these noise sources is a simple matter of transferring the noise from a source, such as a resistor, to the spot in the circuit where you want it. The current noise from a resistor may be measured by attaching the resistor to a zero-level source. Then the current noise would be current flowing through a zero-volt voltage source connected across the resistor (so long as the resistor and voltage source are isolated). Now that the noise current is flowing through a V device, we may use one of the controlled sources, such as an F device, to transfer or "insert" the noise into another part of the circuit.

As an example, let's generate a 1pA/Hertz$^{1/2}$ as our "standard" and create a noise current of 5pA/Hertz$^{1/2}$. At room temperature (298°K) the resistance required for the noise we want is

$$R = \frac{4 \cdot k \cdot T}{\overline{i^2}} = \frac{4 \cdot 1.38 \cdot 10^{-23} \cdot 298}{10^{-12} \cdot 10^{-12}} = 4 \cdot 1.38 \cdot 298 \cdot 10 = 16.45 \, K\Omega \quad \text{(10-4)}$$

Then we connect this resistance across a zero-volt V device, and transfer the noise current with an F device having a gain of 10, to arrive at a noise current level of 10pA/Hertz$^{1/2}$. The circuit file for this might look as shown:

```
Rnoise 1 0 16.45K
Vsense 1 0 DC 0
Fnoise 2 3 Vsense 10
```

Nodes 2 and 3 are the output of the new current noise. And, since it is a noise source, you don't need to worry about which way to connect it into the circuit; just connect it across the input to the operational amplifier.

For a noise voltage source, we could use an analogous technique of measuring the noise voltage across an isolated resistor, then transferring that noise with an

E device. However, we will be lazy by using the same circuit as before. This time we will generate a $1\text{nV/Hertz}^{\frac{1}{2}}$ "standard" and insert a $3\text{nV/Hertz}^{\frac{1}{2}}$ noise. First, let's calculate the resistance required for 1nA (yes, nano-amp, not nano-volt) of noise. Then we use an H device, instead of an F device, which will convert the noise current to a noise voltage. The circuit file for this might look as shown:

```
Rnoise 1 0 16.45m
Vsense 1 0 DC 0
Hnoise 2 3 Vsense 3
```

Nodes 2 and 3 are the output of the new voltage noise. Again, since this is a noise source, you won't need to worry about which way to connect it; just insert the H device in series with the input to the operational amplifier.

These circuits provide the noise levels we need for modeling using "pure" noise (voltage and current) sources so as not to load the circuit at all. This follows the model for referring all noise to the input of the circuit. However, you will need to use a different resistor for each source you make, so that the generated noise is uncorrelated. You **should not use just one resistor** as the basis for all of your noise generators as they would be correlated (note that this is the way to create correlated noise, if that is what you want).

10.7 PARAMETERIZED SUBCIRCUITS FOR NOISE SOURCES

Instead of inserting the few lines each time for a noise source, it would be more productive to invent a subcircuit. This subcircuit can be made more flexible by using a parameter to specify the noise level to be generated. In PSpice, you can define parameters, for use by the subcircuit only, after the node names have been specified. The parameters are separated from the list of node names by the keyword PARAMS: (note the colon, which is required). This enhanced form of the .SUBCKT statement is

.SUBCKT *name node* [[*node...*]] PARAMS: *parameter* [[*parameter...*]]

where each *parameter* definition includes both the name and the default value of the parameter. The default value is used by PSpice when you don't specify any overriding value for the parameter.

For example, a current-noise generator circuit could be defined as

```
.subckt Inoise n1 n2    ; creates current noise
+ params: level=1pA
*
  Rn 1 0 {4*1.38E-23*298/(level*level)}
  Vn 1 0 DC 0
  Fn n1 n2 Vn 1
.ends
```

where the parameter, which is defined on a continuation line for clarity, sets the level of noise generated by the subcircuit. The parameter is then used in a formula for the resistor value, which is the formula for current noise solved for resistance (10-4). Likewise, we could make a voltage-noise generator circuit

```
.subckt Vnoise n1 n2    ; creates voltage noise
+ params: level=1nV
*
  Rn 1 0 {4*1.38E-23*298/(level*level)}
  Vn 1 0 DC 0
  Hn n1 n2 Vn 1
.ends
```

where the only difference is that an H device is used to convert the noise current into a voltage.

EXERCISE 10.7-1

Demonstrate the effects of correlated noise by using the same (generated) noise source twice in a circuit. Re-simulate using two independent noise sources. Were the noise values what you expected?

10.8 NOISE-FREE "RESISTORS"

There is no way, in PSpice, to shut off the noise model for the resistor, but sometimes you may want an element that acts like a resistor without it generating noise. This is easily done by using either the G or H device, which are ideal controlled sources and do not create noise. The key is to see that a resistor can be thought of as a VCCS, where the voltage across the device is proportional to the current through the device, or as a CCVS, where the current through the device is proportional to the voltage across the device. Using the G device, for example:

> *Gname node_1 node_2 (node_1 , node_2) conductance*

the controlling nodes are the same as the device nodes. The value of *conductance* is 1/R, which you will have to calculate and input as a number, since a formula cannot be used for this value (in this form of the G device). Using the H device, for example:

> *Hname node_1 node_2 Vname resistance*
> *Vname node_2 node_3* DC 0

which requires a voltage source to act as a current sensor. As with the G device, the value of *resistance* needs to be calculated and input as a number, since a formula cannot be used for this value (in this form of the H device).

As with most things, there are trade-offs and considerations with using either form. The G device form is simpler and does not create an additional node in the circuit, but it is still an ideal current source and will not provide a DC path. For example, if this form is used in series with a capacitor, the node connecting the "resistor" and capacitor will be strictly considered a floating node. The dual situation occurs for the H device form; if it is used in parallel with an inductor, a voltage-loop is formed. So you cannot automatically use one form for every case.

10.9 CREATING "PINK" AND "FLICKER" NOISE

As mentioned earlier, the thermal noise generated by a resistor is said to be *white noise* in an analogy with human vision: pure white light is made up of equal power from all frequencies. Extending the analogy, colored light has more power from some frequencies than others, so noise with any deviation in its spectral density is said to be *colored* or more specifically called *pink noise*. Traditionally, pink noise is bandlimited white noise as though the noise had been filtered by a low-pass or band-pass frequency-selective circuit. In fact, we can use this approach to create the noise spectra.

The circuit starts simply enough, using the current noise generator circuit covered earlier in this chapter:

```
* Bandlimited (pink) Noise
X1 1 0 Inoise params: level=100nA
G1 1 0 (1,0) 1
L1 1 0 {1/(2*3.14159*10Hz)}
C1 1 0 {1/(2*3.14159*1KHz)}
```

The noise current is impressed across a parallel RLC network. The resistor is formed by a noiseless VCCS, G1, which is set to be the equivalent of 1 Ohm. This converts the noise current into a voltage signal at the same level. The inductor L1 is set by formula to have a reactance of 1 Ohm at 10Hz. At lower frequencies it will shunt the noise current and reduce the noise voltage. Similarly, the capacitor C1 is set by formula to have a reactance of 1 Ohm at 1KHz. At higher frequencies it will shunt the noise current and reduce the noise voltage. The rest of the circuit, not including the definition of Inoise is

```
Vdummy 2 0 DC 0
Rdummy 2 0 1G
.ac dec 100 1Hz 100KHz
.noise v(1) Vdummy 100
.probe
.end
```

The reason for the dummy voltage source is to satisfy the .NOISE statement, and the reason for the dummy resistor is to satisfy the dangling node check in PSpice.

The output noise is shown in Figure 10-7. In normal use, you would transfer this noise with an E device to insert the noise voltage, or use a G device to insert a noise current, in some section of a larger circuit. You might even want to create more subcircuits, with frequency parameters, to generalize this approach.

For example, you might want to create *flicker noise*. This type of noise has a spectral density that decreases proportionally to frequency, so that it has a characteristic 1/f spectrum; for that reason flicker noise is often called *1/f noise*. Nowadays flicker noise is associated with surface recombination current in semiconductor devices; however, it was first discovered in vacuum tubes by J. B. Johnson in 1925. This was interpreted, by W. Schottky in 1926, as fluctuations in the cathode surface causing nonuniform emission. Flicker noise is also known as *contact noise*.

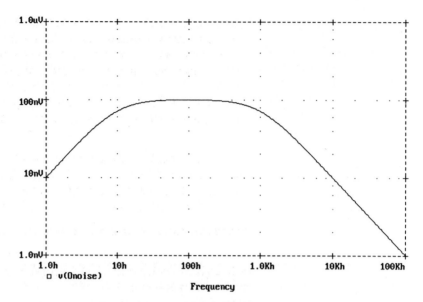

Figure 10-7   Bandlimited (pink) noise.

The relationship of flicker noise to frequency

$$\overline{e^2} \propto \frac{1}{f} \qquad\qquad (10\text{-}5)$$

has been verified experimentally with $1/f$ equalling several months.[†] While this relationship indicates that noise becomes infinite as f approaches 0Hz, the proper way to think about $1/f$ noise is that the noise measured over any range of frequencies will be the same between equal ratios of frequency: From 10Hz to 1Hz, from one second to 10 seconds, from one hour to 10 hours, from one month to 10 months; these measurements will be the same.

We can build on the noise-generating subcircuits constructed earlier to create the flicker noise effect. For example, injecting a noise current into a capacitor will produce a noise voltage that decreases with frequency as the capacitor shunts the noise current. For example:

```
.subckt OFnoise n1 n2  ; creates current noise = k/f
+ params: level=1nA, f0=1KHz
   Xn 1  0 Inoise params: level={level}
   Cn 1  0 {1/(2*3.14159*f0)}
   Rn 1  0 1G
   Gn n1 n2 (1,0) 1
.ends
```

creates current noise that decreases with frequency. This circuit has two parameters, `level` and a "crossover" frequency, `f0`, where the generated noise attains that level.

[†] *Ask EDN*, EDN magazine, page 31, August 19, 1993.

Injecting a current into an ideal capacitor does not provide a DC path for the connecting node, so a large-valued resistor is used to provide a DC path. On its own, this resistor provides less than 10^{-28}amp/Hertz^{½} for any earthly temperature. Also, the corner frequency of the RC combination is at $1/10^9$ the value of f0. This should be a sufficiently low frequency for most purposes.

Similarly, a circuit generating noise proportional to frequency can be constructed. For example:

```
.subckt Fnoise n1 n2    ; creates noise = k· f
+ params: level=1nA, f0=1KHz
   Xn 1  0 inoise params: level={level}
   Ln 1  0 {1/(2*3.14159*f0)}
   Gn n1 n2 (1,0) 1
.ends
```

wherein an inductor is used in place of a capacitor. The parameters to this subcircuit, and their interpretation, are the same as before. Also, all of the DC path considerations are satisfied, so there are no extraneous elements. Such a circuit produces the effect of thermal noise in the gate current of a junction field-effect transistor (JFET): as the real part of the gate impedance decreases, with frequency, the noise increases.

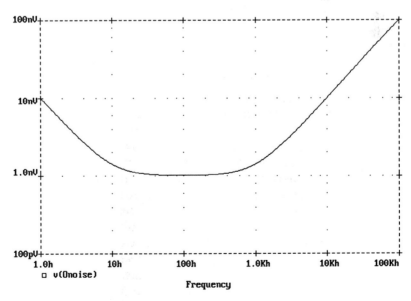

FIGURE 10-8 Thermal, 1/f, and high frequency noise.

We can combine these circuits to form another example that may prove useful in your work:

```
A 1/f plus HiFreq Noise Example
X1 1 0   Inoise params: level=1nA
X3 1 0 OFnoise params: level=1nA, f0=10Hz
X2 1 0   Fnoise params: level=1nA, f0=1KHz
Vsen 1 0 DC 0
Hout 2 0 Vsen 1
Rout 2 0 1G
.ac dec 100 1Hz 100KHz
.noise v(2) Vsen 100
.probe
.end
```

which is missing the various subcircuit definitions. The output noise is shown in Figure 10-8. In normal use, you would transfer this noise, as shown with an H device, to insert the noise voltage, or use an F device to insert a noise current, in some section of a larger circuit.

Transient, or time-domain, response is the most often used analysis for simulators like PSpice. This type of analysis attempts to simulate the operation of your circuit as time progresses and various inputs change level, or as the circuit oscillates (because it is designed to oscillate) under the control of component values. Transient analysis is also the most trouble-prone analysis because of the compromise that needs to be made to either

- take small time-steps to ensure accuracy (but the simulations take a long time to complete), or

- take large time-steps with reduced accuracy (and possibly skip important features of the circuit response).

11.1 SIMULATING TIME

Without getting philosophical about what "time" is, we observe that circuits behave predictably, and repeatably, with the progression of time. The changes in node voltages and branch currents are described by laws, and these descriptions are used by PSpice to simulate how a circuit will behave. We saw this with the DC sweep and small-signal analyses, and it is not much different for transient analysis with the exception that to predict forward in time, the assumption is made that the currents, voltages, and element values will not change much from their present values. This notion is simplified, of course, but as an example you might consider the following: if, at the present time, an amount of current is flowing into a capacitor, then, at the next moment, very much the same amount of current will be flowing. If this is true, the simulator may reliably predict the change in the voltage across the capacitor. If there is a big change in the two amounts of current, then the simulator needs to take smaller time-steps. In the end, if the steps taken were small enough, the calculations will approximate the circuit response for continuous time.

All of this is complicated by the use of active elements (not covered so far, but you have some idea about their operation), which have regions of different operating characteristics. Diodes conduct readily in one direction of current, transistors cease amplifying with small changes in voltage, and so on. These gross changes in operation force the simulator to slow down and step carefully "around the curves" in the simulated response. Sometimes the numerics for doing this fail and produce chaotic results that halt the simulator (non-convergence at a time point). We will look at these problems.

11.2 Specifying Input Sources

You may recall that the independent voltage sources (V device) and current sources (I device) had the statement form

name node node value

where *value* was the DC or AC voltage or current level, depending on the device type. A more complete representation of the input source statement is

name node node [[*DC_value*]] [[*AC_value*]] [[*transient_value*]]

where, if you leave out *DC_value*, the DC value is set to zero, and likewise for *AC_value*. You would include values for all of the situations where you want to use an independent source that has a DC value, an AC value, and a transient value if you were doing all of these types of analyses. The DC value will be used for the operating point analysis and DC sweep. The AC value may combine with the DC value to set the operating point for the small-signal analysis. The transient value will override the other specifications **only** during transient analysis. If *transient_value* is not specified, the DC value will be used and the source is assumed to remain constant during the simulation.

The *transient_value* portion of the statement has several forms, one for each type of waveform. If present, *transient_value* must be one of

EXP *parameters* for an exponential waveform (see page 117)

PULSE *parameters* for a pulse waveform, which may repeat (see page 118)

PWL *parameters* for a piecewise linear waveform (see page 119)

SFFM *parameters* for a frequency-modulated waveform (see page 120)

SIN *parameters* for a sinusoidal waveform (see page 121)

Note: while all of the descriptions are for the independent voltage source, the same parameters are available for the independent current source (except that the units of volts are replaced with amps).

During the transient analysis, all of the independent sources having a transient specification will be activated. The remaining independent sources will maintain the value of their DC specification, or zero if there is no DC specification.

Any of these waveforms may be conveniently set up by using the stimulus editor, "StmEd," which is a Probe-like utility for graphically viewing and editing transient waveform specifications.

General Form

EXP(*V1* *V2* ⟦*Td1* ⟦*Tau1* ⟦*Td2* ⟦*Tau2*⟧⟧⟧⟧)

Example

VRAMP 10 5 EXP(0V .2V 2uS 20us 40uS 20uS)

TABLE 11-1   EXP (exponential) waveform parameters.

EXP **parameters**		Default value	Units
V1	initial voltage	none	volt
V2	peak voltage	none	volt
Td1	rise time delay	0	second
Tau1	rise time constant	*Tstep*	second
Td2	fall time delay	*Td1* + *Tstep*	second
Tau2	fall time constant	*Tstep*	second

Tstep is the print step size specified in the .TRAN statement, which provides sane results if these default values are used. These values are normally specified.

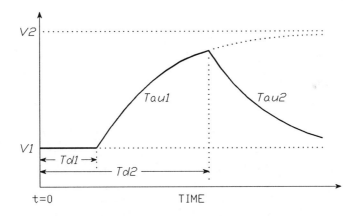

FIGURE 11-1   EXP (exponential) waveform.

The EXP form causes the voltage to be *V1* for the first *Td1* seconds. Then, the voltage decays exponentially from *V1* to *V2* with a time constant of *Tau1*. The decay lasts *Td2*−*Td1* seconds. Then, the voltage decays from *V2* back to *V1* with a time constant of *Tau2*.

General Form

> PULSE(*V1 V2 [[Td [[Tr [[Tf [[Pw [[Period]]]]]]]]*)

Example

> VSW 10 5 PULSE(-1V -2V 50uS .1uS .1us 2uS 10uS)

TABLE 11-2   PULSE waveform parameters.

PULSE **parameters**		Default value	Units
V1	initial voltage	none	volt
V2	peak voltage	none	volt
Td	initial delay time	0	second
Tr	rise time	*Tstep*	second
Tf	fall time	*Tstep*	second
Pw	pulse width	*Tstop*	second
Period	pulse period	*Tstop*	second

Tstep is the print step size and *Tstop* is the stop time that are specified in the .TRAN statement, which provides sane results if these default values are used. These values are normally specified.

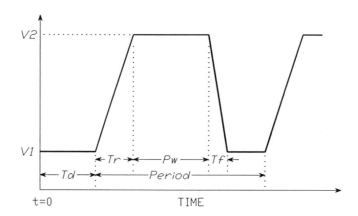

FIGURE 11-2   PULSE waveform.

The PULSE form causes the voltage to start at *V1* and stay there for *Td1* seconds. Then, the voltage goes linearly from *V1* to *V2* during the next *Tr* seconds. Then, the voltage stays at *V2* for *Pw* seconds. Then, the voltage goes linearly from *V2* back to *V1* during the next *Tf* seconds. The voltage stays at *V1* for *Period−Tr−Pw−Tf* seconds, and then the cycle is repeated (except for the initial delay, *Td*).

General Forms

> PWL(*T1 V1* 〚*Tn Vn*〛...)

> PWL 〚TIME_SCALE_FACTOR=*value*〛 〚VALUE_SCALE_FACTOR=*value*〛
> + *corner_point*...

where *corner_point* is one of:
   〚*Tn Vn*〛
   REPEAT FOR *count corner_point*... ENDREPEAT
   REPEAT FOREVER *corner_point*... ENDREPEAT
   FILE *file_name*

Examples

```
V3 10 5 PWL(0,-1V 1uS,0V 10uS,0V 10.1uS,10V 20uS,10V 20.1uS,20V)
V5 17 2 pwl Repeat For  5  1,0  2,1  3,0  EndRepeat
```

TABLE 11-3   PWL (piecewise linear) waveform parameters.

PWL **parameters**		Default value	Units
Tn	time at corner	none	second
Vn	voltage at corner	none	volt

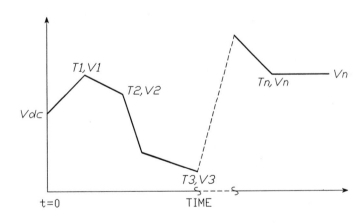

FIGURE 11-3   PWL (piecewise linear) waveform.

The PWL form describes a piecewise linear waveform. Each pair of time-voltage value pairs specifies a corner of the waveform. The voltage at times between corners is the linear interpolation of the voltages at the corners. If the first pair's time is not zero, then the source's DC voltage will be used as the initial value. If the simulation continues beyond the last pair's time, then that pair's voltage will be maintained for the remainder of the simulation.

Please refer to MicroSim's *Circuit Analysis Reference Manual* for further details regarding the *corner_point*-style specification.

General Form

```
SFFM( Vo  Va  [[Fc  [[Mdi  [[Fs]]]] )
```

Example

```
VFM 10 5 SFFM(0 2V 101.1MegHz 5 4KHz)
```

TABLE 11-4   SFFM (single-freq. FM) waveform parameters.

SFFM **parameters**		**Default value**	**Units**
Vo	offset voltage	none	volt
Va	peak amplitude of voltage	none	volt
Fc	carrier frequency	1/*Tstop*	Hertz
Mdi	modulation index	0	
Fs	signal frequency	1/*Tstop*	Hertz

Tstop is the stop time specified in the .TRAN statement, which provides sane results if *Tstop* is used to compute the default values for *Fc* and *Fs*. These values are normally specified.

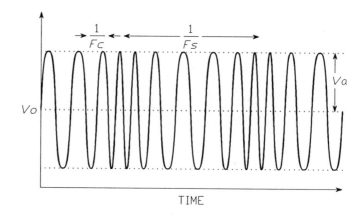

FIGURE 11-4   SFFM (single-freq. FM) waveform.

The SFFM (single-frequency FM) form causes the voltage to follow this formula:

$$Vo + Va \cdot sin\left(2\pi \cdot Fc \cdot time + Mdi \cdot sin(2\pi \cdot Fs \cdot time)\right)$$

General Form

SIN(*Vo* *Va* [[*Freq* [[*Td* [[*Df* [[*Phase*]]]]]])

Example

VSIG 10 5 SIN(0 .01V 100KHz 1mS 1E4 45)

TABLE 11-5   SIN waveform parameters.

SIN **parameters**		**Default value**	**Units**
Vo	offset voltage	none	volt
Va	peak amplitude of voltage	none	volt
Freq	frequency	1/*Tstop*	Hertz
Td	delay time	0	second
Df	damping factor	0	second^{-1}
Phase	phase advance	0	degree

Tstop is the stop time specified in the .TRAN statement, which provides sane results if *Tstop* is used to compute the default value for *Freq*. This value is normally specified.

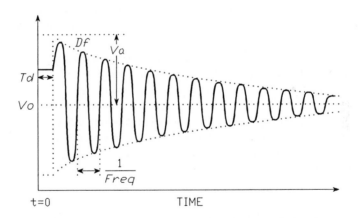

FIGURE 11-5   SIN waveform.

The SIN form causes the voltage to start at *Vo*+*Va*·*sin*(*Phase*·π/180) and stay there for *Td* seconds. Then, the voltage becomes an exponentially-damped sine wave described by this formula:

$$Vo + Va \cdot sin\left(2\pi \cdot \left(Freq \cdot (time - Td) + \frac{Phase}{360}\right)\right) \cdot e^{-(time - Td) \cdot Df}$$

Note: the SIN waveform is for transient analysis **only**. It does **not** have any effect during small-signal (.AC) analysis, which is a common mistake. To give a voltage a value during small-signal analysis use an AC specification; for example:

```
VAC   3   0   AC 1V
```

will have an amplitude of 1 volt during small-signal analysis and be zero during transient analysis, whereas

```
VTRAN   3   0   SIN(0 1V 1KHz)
```

will be the other way around.

11.3 THE .TRAN STATEMENT

The .TRAN statement specifies the time interval over which the transient analysis takes place. It also specifies some limits on the way PSpice performs the analysis and when hard-copy output will be generated. The statement form is

.TRAN[[/OP]] *print_interval final_time* [[*no-print_interval* [[*step_ceiling*]]]]

For the simulator, "time" always starts at zero and proceeds up to the value of *final_time* (a.k.a. *Tstop*). The /OP option, which stands for OPERATING POINT, commands PSpice to print out the table of node voltages calculated from the bias-point calculation for the transient analysis. Normally these voltages would be the same as the bias-point calculation from the other analyses, such as frequency analysis, unless you specified some initial conditions that apply only to transient analysis. Then the bias-point is likely to be different, and /OP will save the node voltage values in the output file.

The value for *print_interval* specifies when hard-copy (.PRINT and .PLOT) output will be generated. The value for *no-print_interval* will suspend both hard-copy and Probe output until that amount of simulated time passes, so you will only have these output(s) for the final stretch of the simulation.

The value for *step_ceiling* specifies the maximum size of time step PSpice may take in working through the transient simulation. In general, if *step_ceiling* is not specified, PSpice uses *final_time*/50, which is also the maximum time-step size for any transient simulation.

11.4 PRINT AND PLOT OUTPUT

Output from transient analysis may be generated by .PRINT or .PLOT statements, just as in DC and small-signal analysis. In either case, the output is organized by the time at which the calculations were made. The statement forms are

.PRINT TRAN *output...*

and

.PLOT TRAN *output...*

Each *output* entry becomes a column in the table output by the .PRINT statement, or a curve in the plot output by the .PLOT statement. The output values you can print/plot are node voltages and device currents (which also means source currents, as a source is also a device). For example:

```
.PRINT TRAN V(7)
```
prints the voltage at node 7

```
.PRINT TRAN I(R1)
```
prints the current through R1

You may print several values in one table, and mix voltages and currents; for example:

```
.PRINT TRAN V(3) I(R2)
```

Usually you will want to print the analysis time in the first column to simplify finding results in the table, so PSpice does this for you; the first column, which comes before the columns you specify, always contains the analysis time.

11.5 GRAPHICS OUTPUT AND CALCULATIONS

Using Probe with transient analysis is identical to what we have done before with DC and small-signal analysis; just include a `.PROBE` statement in the circuit file. Let's try simulating our LC-filter example with a time-varying stimulus. Initially, we will use a step waveform to simulate overshoot and ringing. Recalling the LC-filter circuit from the frequency analysis section:

```
Four double-pole, low-pass, LC-filters
Vin 1 0 pwl(0,0 .1m,1 5m,1 5.1m,0)
*   Q = .5
R1  1  2  200
L1  2  3  10mH
C1  3  0  1uF
*   Q = 1
R2  1  4  100
L2  4  5  10mH
c2  5  0  1uF
*   Q = 2
R3  1  6  50
L3  6  7  10mH
C3  7  0  1uF
*   Q = 4
R4  1  8  25
L4  8  9  10mH
C4  9  0  1uF
.tran 1m 10m
.probe
.end
```

You can see that a piecewise-linear source was specified for the step input. This step has a 0.1-millisecond transition and a nearly 5-millisecond duration. The transient analysis has been specified to last for 10 milliseconds.

Looking at the response of the filter sections, in Figure 11-6, we see overshoot and ringing in the higher-Q filters. Besides just looking at voltages and currents, with Probe, we can also make measurements using formulas.

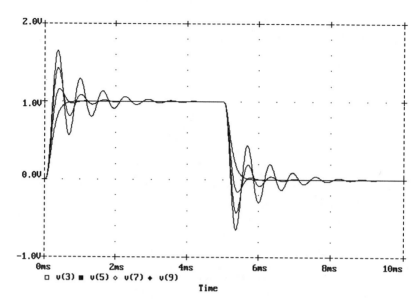

FIGURE 11-6 Step excitation of LC filter.

In Figure 11-7, we see two derived measurements for

- the volt-amp product, or instantaneous power dissipation, of resistor R4, and

- the running RMS average of this volt-amp product, or average power dissipation.

If the step function were continued, as an oscillation, the RMS average across one cycle would represent the long-term power dissipation of this resistor. In this case the dissipation would be slightly less than 0.3 milliwatts.

By using sinusoidal excitation, we can also see effects that were demonstrated using frequency analysis. For sinusoidal excitation we modify the circuit file for the input source, as shown:

```
Vin 1 0 sin(0 1 2000)
```

which becomes a 2,000 Hertz sine wave with a 1-volt peak amplitude.

In Figure 11-8, we see different amplitudes of response to the same input. Since the input sine wave is slightly above the resonant frequency of the filter sections, the amplitude of response is set by the Q of the sections, with the higher-Q sections having greater response. Also, the higher-Q sections have greater phase lag. If you look closely, you will notice that the zero crossings of the higher-Q sections come after those for the lower-Q sections.

Finally, the step overshoot we saw previously is also apparent in the transient response to sinusoidal input. This initial transient is present because there is a change in the input, so that at TIME=0 the input changes from nothing to a sine

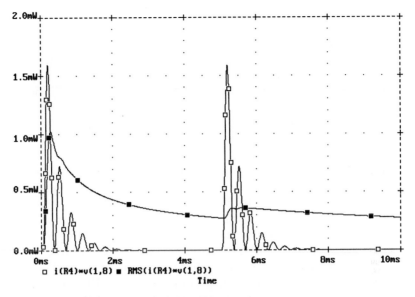

FIGURE 11-7 Plot of instantaneous and average power.

wave. This is a step change, and there is an overshoot and ringing period that dies out to become the steady-state response. This is one difference between transient and frequency response.

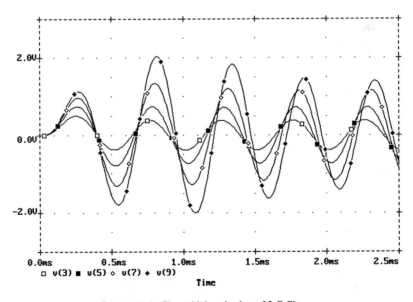

FIGURE 11-8 Sinusoidal excitation of LC filter.

11.6 SETTING INITIAL CONDITIONS

For many of the simulations you run, you will want the circuit to begin with particular node voltages or device currents; for example, you may want to start with a particular inductor current. You need to set the "initial conditions" for the simulation. There are two ways of setting initial conditions, and a related way of influencing the initial condition (for added confusion, which we will try to correct).

The first statement that directly sets the initial conditions is

> .IC V(*node* [[, *node*]]) = *voltage...*

which presents a list of nodes, or node pairs, to be biased to the indicated voltage. These statements are effective for both transient **and** small-signal analyses, the latter being useful for analyzing the frequency response of systems in a non-quiescent state.

During the bias-point calculations, which is the setup for the transient analysis, PSpice connects voltage sources, each having a source resistance of 0.002 Ohm, to the nodes specified by the .IC statements in your circuit file. This effectively fixes these nodes to those voltage levels. Also, the source resistance protects against voltage loops that are usually impossible to resolve. Then, just before the transient analysis begins, the voltage sources are removed and the circuit is initialized. The same "apply-then-remove" sequence is used during the bias-point calculations for small-signal analysis.

The second technique that sets initial conditions is used for devices instead of nodes. In this case, it is not a separate statement but a part of the capacitor and inductor component statement. This augmentation of the statement was kept from you, until now, so you can see how it fits with the other methods of setting initial conditions. For the capacitor, you may set the initial voltage impressed on the plates:

> *<capacitor statement>* IC= *initial_voltage*

For the inductor, you may set the initial current flowing through the windings:

> *<inductor statement>* IC= *initial_current*

These settings, unlike the .IC statement, are used for transient analysis **only**. Furthermore, the IC= option is used **only** when the .TRAN statement includes the UIC option, which stands for <u>U</u>SE <u>IC</u>= , as shown:

> .TRAN [[/OP]] *<list of time-values>* [[UIC]]

This commands PSpice to skip the bias-point calculations and proceed to the transient analysis.

Setting the initial capacitor voltage this way lets you specify the relative voltages without having to specify the referenced-to-ground-voltages of the nodes of the capacitor. For the inductor, this is the only way to easily specify the current in the windings. Again, you do not need to specify the referenced-to-ground-voltages at the nodes of the inductor.

Of course, you can **over specify** the initial conditions by using both .IC and IC= causing conflicting voltage levels across a capacitor. Be careful.

The third, related way of setting up the simulation works during the bias-point calculation of all the analyses. This is the `.NODESET` statement, which is similar to the `.IC` statement, in form:

> `.NODESET V( node [, node]] )` = *voltage...*

It does **not** force the initial voltage at a node. Rather, it provides PSpice with an initial guess at the outcome of the bias-point calculation and operates as follows:

- voltage sources with source resistance of 0.002 Ohm (like the ones used for the `.IC` statement) are connected to the circuit where specified, then

- the bias-point calculations are made and the circuit converges to a set of node voltages, then

- the `.NODESET` voltage sources, inserted in the first step, are removed, "releasing" the circuit, then

- the bias-point calculations continue without the `.NODESET` voltage sources connected, and the final bias-point voltages are calculated.

As you can see, the `.NODESET` statement compels PSpice to use the specified voltages for only the first half of the bias-point calculation; then PSpice is free to recalculate the node voltages. This technique lets you give "hints" to PSpice about the bias-point voltages without fixing the nodes to a particular voltage. This is particularly useful for circuits with more than one stable solution, especially balanced circuits (for example, flip-flops), which may have a meta-stable state that PSpice can easily use as the bias-point. The "hints" provided by the `.NODESET` statement may not even need to be that accurate, so long as they decide the issue of how you want the circuit to be biased.

To repeat, the `.NODESET` statement is used during the bias-point calculation for all of the analyses. Of course, if you have both statements, `.IC` and `.NODESET`, specified for the same node, the .IC statement will override during the bias-point calculation for transient analysis **only**.

11.7 HAZARDS: PROBLEMS OF TIME-STEPPED SOLUTIONS

The major problem of transient analysis is accuracy or, to be pessimistic, error. Since the circuit equations are solved numerically instead of analytically, approximations are used to extrapolate circuit operation at the next instant in time. Accuracy becomes a question of how good the approximations are and how far their extrapolations may be trusted (see Appendix D.4 for references explaining the operation of SPICE). Simulation accuracy is controlled by the parameter settings for `RELTOL`, `VNTOL`, `ABSTOL`, and `CHGTOL`, which are set by using the `.OPTIONS` statement (see Appendix B). The trade-off is computation time.

The foremost of the parameters is `RELTOL`, which sets the relative accuracy of the calculated voltages and currents. `RELTOL` is the numerical ratio of error allowed to the signal level; for example, a RELTOL value of 0.01 means that the voltages and currents are to be calculated to within 1% of their "real" values. But, how does the simulator determine what the "real" value is without calculating it? It doesn't... but the mathematical properties of the solution method let you estimate how close you are to the "real" value. So the "real" value is within RELTOL of the answer calculated by PSpice. Picture a "band of uncertainty," with a width that is $2 \times RELTOL \times value$, centered on the waveform calculated by PSpice. Somewhere, in this band, lies the "real" waveform.

The other tolerance parameters, VNTOL, ABSTOL, and CHGTOL, set the best accuracy for the **voltages**, **currents**, and capacitor **charges** (and inductor **fluxes**), respectively. That is, these parameters set the least error, or, optimistically, the most accuracy, allowed in terms of an absolute amount. Now, why do we need this? Think about a waveform with a value that changes sign. As it approaches zero, maintaining RELTOL accuracy will force the simulator to work harder because the tolerance, in absolute terms, is getting tighter. Ultimately, at zero, the uncertainty band containing the "real" waveform has zero width, and there is no assurance that the "real" value could ever be calculated. These tolerance parameters set the minimum error allowed, and the uncertainty band's width becomes, for a voltage, twice the maximum of $RELTOL \times value$ and VNTOL.

As the transient solution progresses, internal calculations are made for the next time point to evaluate the circuit. The size of each time-step is set as the minimum of several factors, but usually it is being set by a calculation of the errors involved in the integration techniques used for the energy storage elements (such as capacitors) in the circuit. If the node voltages are changing rapidly, small time-steps are used to calculate accurately "around the curves." If the circuit becomes less active, the node voltages are more stable and larger time-steps may be used.

Using ever larger time-steps becomes a problem when the step size exceeds the Nyquist rate for the signals in your circuit. After all, the simulator is like a sampled-data system with samples being made at every time-step. If the samples become too widely spaced, high-frequency operation will be *aliased* to a lower frequency. For example, an astable multivibrator circuit may be incorrectly simulated by allowing the simulator to have a time-step size that is larger than one-half the oscillation period. In this case, the standard capacitor charge/discharge circuitry is never activated because the simulator has found a "stable" solution (a lie) where the discharge threshold is never reached.

To prevent the time-step from becoming too large and missing changes in the circuit, PSpice forces a time-step at each corner in the driving waveforms; for example, the PULSE and PWL specifications. PSpice also limits the time-step to 1/8th of the cycle time of the sinusoidal source with the highest frequency. Another limit is twice the previous time-step size, which has proven to be a good, conservative measure for not allowing the simulator to get too far ahead and into trouble. In addition, you may also set the maximum step-size by using the fourth parameter of the .TRAN statement. PSpice uses the minimum of all the limits described to limit the size of the time-step.

11.8 BENEFITS: TRANSIENT SOLUTIONS FOR STATIC PROBLEMS

Transient analysis can provide some solutions to problems that you would ordinarily consider as static problems. The classic example is analyzing the transfer characteristics of a circuit with regenerative feedback, such as a Schmitt trigger. These types of circuits cannot be analyzed easily using a DC sweep. The circuit has a region with two stable operating points, and the simulator will need to jump discontinuously from one stable solution to the other at each end of this region.

By using transient analysis and the piecewise linear source, you can apply a slowly varying ramp to the same circuit, just as you might with the DC sweep. The difference here is that the circuit becomes discontinuous only in the sense of its DC operating point; in transient analysis the circuit is continuous, although at some threshold it does transfer rapidly from one operating point to the other. When the circuit switches like this, PSpice reduces its time-step size to carefully analyze the transition. During the remainder of the time, the time-steps become quite large because the node voltages do not change so much.

This kind of treatment duplicates exactly what you would do on the workbench; you would slowly change the input and measure the output. Also, you would probably sweep the input in one direction, until the circuit switches, and then sweep in the other direction, until the circuit switches back. Doing this measures the hysteresis of the circuit. This is easy to do with the piecewise linear source in one transient run, but with DC analysis you would need to do a run for each direction of the sweep.

We can also use a rapidly varying ramp to measure the size of energy storage elements in a circuit, or device. As we reviewed earlier, a linear capacitor stores a charge according to the equation $Q = C \cdot V$. Differentiating this equation with respect to time, we get

$$\frac{\partial Q}{\partial t} = I = C \frac{dV}{dt} + V \frac{dC}{dt} \qquad (11\text{-}1)$$

which simplifies to

$$I = C \frac{dV}{dt} \qquad (11\text{-}2)$$

since:

- if the capacitor is linear, dC is zero, and

- if the capacitor is nonlinear, the original equation is wrong. Charge is the integral of the capacitance, with respect to voltage; using integration by parts, with both voltage and capacitance as functions of time, before differentiating (the equation with respect to time) leads to the same result.

Minor rearranging yields

$$C = \frac{I}{\left(\dfrac{dV}{dt}\right)} \qquad (11\text{-}3)$$

And, of course, similar analysis for inductors will yield the equation

$$L = \frac{V}{\left(\dfrac{dV}{dt}\right)} \qquad (11\text{-}4)$$

Let's get back to transient analysis techniques. We now have a way to measure the capacitance of a network versus voltage. By applying a voltage ramp that produces currents, due to capacitance, that are much larger than the DC currents (for the same voltage levels), the current represents the value of the capacitance along the ramp. Or, we could measure inductance versus current by using a current ramp. Just be sure to isolate and measure one energy storage element at a time. (Later in the book we will see this technique used to plot the capacitance versus voltage characteristics of the semiconductor devices.)

11.9 UNUSUAL WAVEFORM SOURCES

Sometimes your simulations will need a stimulus that is more exotic than a simple sine wave or piecewise linear function. With some ingenuity, you will find that you can create many unusual waveforms by combining the ones available through the use of controlled sources. For example, as shown in Figure 11-9, to have a 10KHz signal riding on a 60Hz power line you might use the voltage-controlled voltage source:

```
vsig   1 0 sin(0 1 10k)
rsig   1 0 1
vpwr   2 0 sin(0 120 60)
rpwr   2 0 1
eboth  3 0 poly(2) (1,0) (2,0) 0 1 1
```

although you could have added these signals by putting the sources in series.

Multiplying, or modulating, signals is another way to generate desired inputs. For example, as shown in Figure 11-10, to create a ten-cycle burst of sine waves you might multiply a sine wave source with a piecewise "switching" function:

```
vsig   1 0 sin(0 1 1K)
rsig   1 0 1
vsw    2 0 pwl(0,0 .001m,1 9.999m,1 10m,0)
rsw    2 0 1
eboth  3 0 poly(2) (1,0) (2,0) 0 0 0 0 1
```

which will produce ten cycles of a 1KHz sine wave at the start of the simulation, and then revert to zero for the remainder of the run. This might be used to simulate the reaction of a filter circuit. You may notice that, in this example, a 0.001-millisecond transition time was used for the switching function. This is a much shorter time than the 1-millisecond cycle time of the sine wave, so it should not deviate much from a

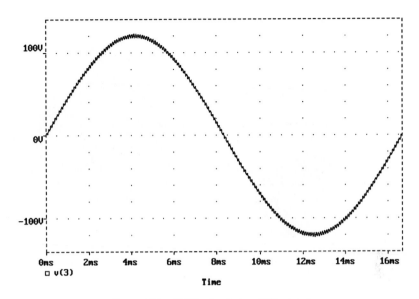

FIGURE 11-9 10KHz signal atop 60Hz signal.

true sinusoidal shape. However, the transition should not be too short as PSpice will cut back on its time-step to process the changes in the switching function. Then it will take some time to get moving again with the simulation.

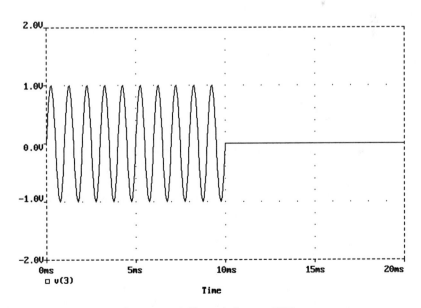

FIGURE 11-10 Ten-cycle burst at 1KHz.

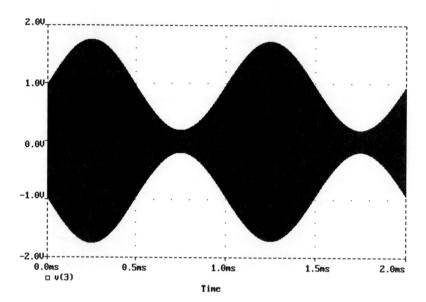

FIGURE 11-11   80% modulation of 500KHz carrier.

By extension, you could build several of these generators with interleaved "on" times to step through a series of frequencies. You have to be careful that the zero crossings of the sine waves occur when the switching functions transfer from one source to another, to get a clean transition. You may need to adjust the initial phase of a source to make this happen. Of course, it is a good idea to do a run with just the sources to check their operation.

Amplitude modulation may be accomplished in a similar way. For example, as shown in Figure 11-11, modulating a 500KHz signal with a 1KHz sine wave, using an 80% modulation index:

```
vsig  1 0 sin(0 1 500K)
rsig  1 0 1
vmod  2 0 sin(1 .8 1K)
rmod  2 0 1
eboth 3 0 poly(2) (1,0) (2,0) 0 0 0 0 1
```

Since we are performing a simple multiplication of the two signals, the modulation index is the ratio of the peak amplitude of the modulating signal to unity. For the same reason, the modulating signal has an offset of 1 volt, or unity, so that if the carrier signal were unmodulated its peak amplitude would be 1 volt.

CHAPTER 12
DISTORTION AND SPECTRAL ANALYSIS

Signal distortions come in many forms (most of them undesirable) and are usually the product of nonlinearity in the gain, or nonuniformity in the phase, of a circuit. The most common types of distortion have been categorized and named:

- *Harmonic distortion* comes from nonlinear gain. The output of the circuit contains integral multiples, or harmonics, of the fundamental input frequencies.

- *Phase distortion* comes from nonlinear phase versus frequency response. This gives rise to "echoes" in the output that precede and follow the main response, resulting in a distortion of the output signal when the input signal has many frequency components.

- *Intermodulation distortion* comes from mixing signals at different frequencies. The output of the circuit contains signals at integral multiples of the sum or difference of the original frequencies.

- *Cross-modulation distortion* occurs when the modulation of one signal is unintentionally transferred to another signal in the circuit.

- *Crossover distortion* comes from nonlinearities in amplification as the signal crosses over between regions of amplifier operation (such as a "push-pull" amplifier).

12.1 THE .DISTO ANALYSIS

The SPICE2 simulator, from U.C. Berkeley, introduced a type of analysis called .DISTO for evaluating some distortion measures. This analysis is performed in conjunction with a frequency analysis, the way noise analysis is done: that is, calculations are performed at the frequencies specified by the .AC statement at the time the frequency analysis is done. The .DISTO analysis calculates the magnitude of power into a load resistor for the following small-signal harmonic products:

the power at $2 \times freq$

the power at $3 \times freq$

where *freq* is each frequency specified as part of the .AC statement, and for the following small-signal intermodulation products:

the power at *freq* + k × *freq*

the power at *freq* − k × *freq*

the power at 2 × *freq* − k × *freq*

where the factor k is specified by the .DISTO statement. The relative magnitude of the second signal, represented by k × *freq*, may also be set by the .DISTO statement. The results of the .DISTO analysis are available to the .PRINT and .PLOT output statements.

PSpice does not include the SPICE2 .DISTO calculations because:

- The calculations are for small-signal distortion **only**, whereas many of the interesting distortion analyses are for large signals.

- The calculations are for power at only a few, selected frequencies, whereas most circuits exhibit distortion components at a large number of frequencies.

- Moreover, the results from SPICE2 were incorrect, particularly for MOS circuits, where the small-signal (analytic) model had not kept up with advances in the large-signal (nonlinear) model.

Instead, PSpice makes use of spectral analysis techniques to calculate distortion. (Note: the recent versions of SPICE3 include a revised, non-analytic technique for calculating distortion.)

12.2 Harmonic (Fourier) Decomposition

One type of spectral analysis, which is performed **in conjunction** with the transient analysis, is called *Fourier analysis*. It is named after Jean Fourier (1768−1830), who demonstrated that any periodic function could be expressed as the sum, or series, of sinusoidal functions. Moreover, if a periodic function is expressed this way, each sinusoid, or *component*, in the series must be periodic over the same interval as the original function. This happens only with sinusoids having frequencies that are an integer multiple of the frequency of the original function; that is, the sinusoids are *harmonics* of the original function's *fundamental* frequency.

The .FOUR statement directs the simulator to perform a *harmonic decomposition*, calculating the *Fourier coefficients* for the sinusoidal components of any voltage or current that you could .PRINT or .PLOT. Also, you select the fundamental frequency on which to base the decomposition. These calculations create tabulated results that include the DC component, the fundamental component, and the components of the second through ninth harmonic of the fundamental (by default). The magnitude and phase values, both absolute and relative to the fundamental, are compiled in the table. Finally, harmonic distortion is calculated.

It is important to remember that when a harmonic decomposition of a transient waveform is taken, **only part** of the waveform is used for the decomposition. The period of time used is the inverse of the frequency, or one cycle's time, of the fundamental frequency you specify for the decomposition. The segment of the waveform that is used is the last period (1/*freq*) of the transient simulation. You will want to set up your simulations so that the segment of waveform decomposed is at the end of the simulation (usually you will just set the time limit for the entire transient run to be one cycle's time of the fundamental).

Finally, a note about harmonic distortion, which is defined to be the ratio of the root-mean-square (RMS) sum of the magnitude of the harmonics to the magnitude of the fundamental, or as a percentage:

$$\% \text{ harmonic distortion} = 100 \times \sqrt{\frac{V_2^2 + V_3^2 + V_4^2 + \cdots V_n^2}{V_1^2}} \qquad (12\text{-}1)$$

where V_n^2 is the magnitude of the *n*th harmonic, squared. Calculating the popular distortion measure called *percent total harmonic distortion* (%THD) is straightforward, and is done for you by PSpice. Also, while setting up the transient analysis you can select the signal level, for example, to include the effects of crossover and clipping.

12.3 THE .FOUR STATEMENT

Remember, you have to be performing a transient simulation to get a harmonic decomposition using the .FOUR statement. Having covered that, we now look at the .FOUR statement and find that it is similar to a .PRINT statement, with the form

.FOUR *fundamental_frequency output_value...*

One, or more, node voltages and/or device currents may be selected for harmonic decomposition. You have to choose the fundamental frequency for this analysis. Let's take a look at a small circuit that demonstrates Fourier analysis:

```
Fourier decomposition
vin 1 0 sin(0 .57 1000)
rin 1 0 1G
e3   2 0 poly(1) (1,0) 0 1 0 -1
r3   2 0 1G
.tran 1u 1m
.four 1000 v(1) v(2)
.probe
.end
```

This circuit performs a transient analysis of a 1KHz sine wave exciting a VCVS with the transfer function $x + x^3$. This is a cubic polynomial that was selected to demonstrate the distortion effects of nonlinear gain and "soft" clipping, since the input waveform has a peak value of nearly $1/3^{1/2}$, which is the point at which the cubic polynomial reverses direction. The transient simulation is run for one cycle of the input sine wave, 1 millisecond, since this is the fundamental waveform of the simulation. Note that both the input and output voltages were selected for decomposition. Seeing the decomposition of a known waveform will provide some guidance in interpreting the results.

The .PROBE statement was included so that we might look at the waveforms; it is not required, and neither is any other output statement, like .PRINT or .PLOT, for the Fourier decomposition.

12.4 LARGE-SIGNAL DISTORTION

Running the circuit described previously yields results not obtainable using the .DISTO analysis; since the DC bias-point has the input voltage at zero, the linear term of the transfer polynomial will dominate and the output will show very little harmonic distortion. Looking at the output file, we find two tables. The first is for the input voltage, V(1):

```
FOURIER COMPONENTS OF TRANSIENT RESPONSE V(1)

DC COMPONENT =   2.017000E-10
```

HARMONIC NO	FREQUENCY (HZ)	FOURIER COMPONENT	NORMALIZED COMPONENT	PHASE (DEG)	NORMALIZED PHASE (DEG)
1	1.000E+03	5.700E-01	1.000E+00	3.062E-07	0.000E+00
2	2.000E+03	3.719E-10	6.525E-10	1.297E+02	1.297E+02
3	3.000E+03	1.568E-09	2.750E-09	8.906E+01	8.906E+01
4	4.000E+03	5.704E-10	1.001E-09	1.554E+02	1.554E+02
5	5.000E+03	9.842E-10	1.727E-09	-2.179E+01	-2.179E+01
6	6.000E+03	5.810E-10	1.019E-09	1.216E+02	1.216E+02
7	7.000E+03	2.297E-09	4.030E-09	1.311E+02	1.311E+02
8	8.000E+03	7.364E-10	1.292E-09	3.312E+01	3.312E+01
9	9.000E+03	9.024E-10	1.583E-09	-1.565E+02	-1.565E+02

```
TOTAL HARMONIC DISTORTION =   5.781659E-07 PERCENT
```

and another table for the output voltage, V(2):

```
FOURIER COMPONENTS OF TRANSIENT RESPONSE V(2)

DC COMPONENT =   1.790805E-10
```

HARMONIC NO	FREQUENCY (HZ)	FOURIER COMPONENT	NORMALIZED COMPONENT	PHASE (DEG)	NORMALIZED PHASE (DEG)
1	1.000E+03	4.311E-01	1.000E+00	3.707E-07	0.000E+00
2	2.000E+03	4.006E-10	9.292E-10	1.282E+02	1.282E+02
3	3.000E+03	4.630E-02	1.074E-01	7.127E-06	6.756E-06
4	4.000E+03	5.130E-10	1.190E-09	1.408E+02	1.408E+02
5	5.000E+03	1.374E-09	3.187E-09	1.273E+02	1.273E+02
6	6.000E+03	4.424E-10	1.026E-09	1.073E+02	1.073E+02
7	7.000E+03	1.007E-10	2.336E-10	1.263E+02	1.263E+02
8	8.000E+03	5.842E-10	1.355E-09	4.881E+01	4.881E+01
9	9.000E+03	1.106E-09	2.566E-09	1.530E+02	1.530E+02

```
TOTAL HARMONIC DISTORTION =   1.073909E+01 PERCENT
```

For each of the selected voltages, the tables show the magnitude and phase of the components at $2\times$, $3\times$, $\cdots 9\times$ the fundamental frequency. These are the harmonics of the fundamental. The values are also shown in a normalized format so that the magnitudes are normalized with the fundamental at unity, and the phases are normalized with the fundamental at zero.

Above the table, the DC component of the waveform is shown. Below the table, the calculation for *total harmonic distortion* is shown (by default, the calculation uses only the second through ninth harmonics).

Looking at the decomposition table for the input sine wave, we can see the limits of the calculations. This is a sort of "noise floor" due to the numerics. The input ideally has a zero DC level, and zero magnitude for all harmonics resulting in zero harmonic distortion. Instead, PSpice arrived at values of around one part in a billion relative to the fundamental. This is numerical noise. From this we can see that only those values that are larger than this "noise floor" are the significant (and useful) values.

The same analysis applies to the phase values. As harmonic magnitudes become small it becomes difficult to determine phase. Again, this is a numerical noise problem. For simple systems, where the output is a periodic waveform, you know from Fourier analysis that the harmonics are either in phase (0°) or out of phase (180°). So, sometimes you have to apply some judgment in reading the tables.

Looking at the decomposition table for the output waveform, we can see that there is one major harmonic at three times the frequency of the fundamental. It is in phase and has a level of about 11% of the fundamental. Accordingly, the total harmonic distortion is also calculated at about 11%.

EXERCISE 12.4-1

Show that the `.FOUR` results will yield smaller distortion values by reducing the size of the input sine wave. Try a 0.1-volt peak and a 0.01-volt peak.

12.5 HARMONIC RECOMPOSITION

One of the uses of the decomposition table printed by the Fourier analysis is providing the harmonic information for regenerating the signal that was decomposed. This may provide you with a compact and fast technique for packaging a complex waveform. Of course, it needs to be a periodic waveform that can be represented adequately with the first nine harmonics.

Looking at the previous table for the cubic transfer function, we find that the output waveform can be represented by summing two sine waves: one is the fundamental, with a magnitude of 0.4311; the other is at $3\times$ the fundamental frequency, with a magnitude of 0.0463. We can combine these signals in the original circuit and compare our "recomposed" signal to the original. This is done by adding two current sources, which sum to develop a voltage across a 1-Ohm resistor, to the

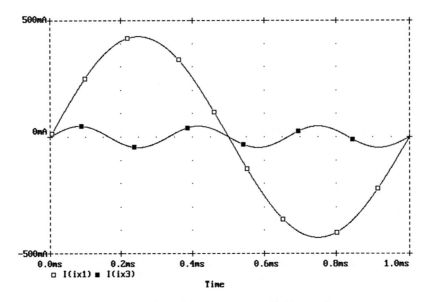

FIGURE 12-1 Plot of fundamental and third harmonic.

circuit:

```
Fourier decomposition
vin 1 0 sin(0 .57 1000)
rin 1 0 1G
e3  2 0 poly(1) (1,0) 0 1 0 -1
r3  2 0 1G
ix1 0 3 sin(0 .4311 1000)
ix3 0 3 sin(0 .0463 3000)
rx  3 0 1
.tran 1u 1m
.four 1000 v(1) v(2) v(3)
.probe
.end
```

Looking at the output with Probe (see Figure 12-1) we can see the components of our "recomposed" waveform. We can also check the accuracy of our work by checking the difference between the original waveform and the "recomposed" signal. This is done in Figure 12-2. Notice the scale for the Y-axis, showing that the difference is quite small relative to the magnitude of the waveforms being compared.

EXERCISE 12.5-1

Do the same analysis for a circuit with the polynomial function $x - x^5$. What maximum input level will you use to show "soft" clipping? Look at the harmonics and compare their magnitudes with the ones in the example.

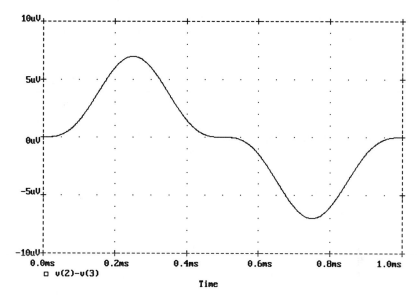

FIGURE 12-2 Plot of error in recomposed signal.

12.6 FOURIER TRANSFORM

Another type of Fourier analysis is part of Probe, which has the capability to calculate a *Fourier transform* of the data sequences in the Probe data file. While the Fourier transform is a special case of the Laplace transform, the Fourier integral (the function performing the transform) can also be thought of as an extension of the Fourier series we saw in the previous section; by extending the fundamental period to infinity each harmonic component becomes infinitesimally close in "frequency." Thus, in the limit, the Fourier series becomes the Fourier integral. This means the Fourier transform converts a function of time to a function of frequency, and vice versa. The physical interpretation of Fourier transform is the conversion of a time-domain signal to the AC steady-state frequency content, or spectrum, which makes up the signal. However, the inverse Fourier transform converts the AC steady-state response, or gain, of a system into the time-domain response of that system to a stimulus with a flat input spectrum, which is an impulse.

The Fourier transform in Probe is a *discrete Fourier transform* (DFT), where the Fourier integral has been replaced by a nearly equivalent summation formula applied to evenly spaced samples of the signal. Furthermore, the transform is accomplished by a special technique credited to Cooley and Tukey, which is commonly called a *fast Fourier transform* (FFT). This is a numerical trick whereby if the size of the data sequence is a power of two (such as 256, or 4096) a much shorter sequence of calculations can be used to get the same results as the discrete Fourier transform. Even for modest data sets the DFT is so time consuming that most computer applications use the FFT (for example, for 1024 points only about 1% of the compute time is required using the FFT versus the "brute force" DFT). Probe uses the FFT, and gets its data by interpolating the data file values to get a set of values with the appropriate number of data points. Then, the FFT is performed.

For Probe, the Fourier transform is a mode; all displayed values or formulas are transformed before being displayed (we will use the word "signal" to also mean any Probe formula of signals). As you might expect, there are some items to be cautious about when using Fourier transforms:

- If the signal is non-zero for a finite interval, you should transform that entire interval. Alternatively, if the signal goes on forever, transform an interval that is typical of what the signal looks like at the other times.

- The transform should be made on a *band-limited* signal. The usual warnings about data sampling at greater than the Nyquist critical rate are taken care of by PSpice and Probe, since the simulation itself is subject to the Nyquist sampling theorem and the time-steps in the transient simulation are usually much smaller than the highest frequency signal. Probe uses the total number of time-steps to guide the FFT. However, if there is a difference in the value between the beginning and end of the waveform, this is a discontinuity that implies a DC level and/or a high-frequency content that will be *aliased* (frequency shifted) into the result from the FFT. Fortunately, this will be evident when you look at the transformed signal, so you may need to retry the simulation for a different interval.

Let's use a variation on the circuit for the Fourier decomposition example to demonstrate Fourier transforms in Probe:

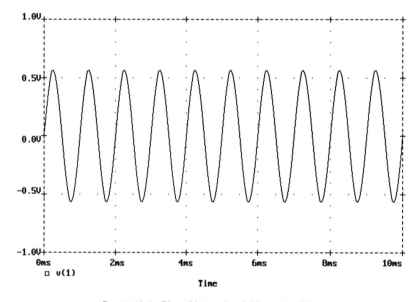

FIGURE 12-3 Plot of input signal (time domain).

```
Fourier transform example
i1  0 1 sin(0 .57 1000)
rin 1 0 1
e3   2 0 poly(1) (1,0) 0 1 0 -1
r3   2 0 1G
.tran 50u 10m 0 50u
.probe
.end
```

This circuit is the same as before, except that a current source is used to create the input voltage. This will make it easy to add in other signals, say, to show intermodulation effects. The transient simulation is run for 10 milliseconds, or 10 cycles, so that the frequency resolution will be 1/10 of the fundamental frequency. The time step is limited to 50 microseconds, or 1/20 of the cycle time, so that the frequency band, from zero up to the Nyquist critical frequency, will extend to about 10 times the fundamental frequency (actually set by the number of data points selected by Probe. This will always be at least as large as the number of points in the waveform data). By running the simulation and then looking at the results with Probe (see Figure 12-3), we see the input voltage in the time domain.

By selecting the X-axis menu and Fourier menu item, we can look at (after some delay for calculation) the input signal's spectrum in *volts-per-root-Hertz* (V/Hz$^{1/2}$), since the input signal is in volts. See Figure 12-4.

Some comments are required here to help you interpret the results of the transform:

• The total power (or mean-square amplitude) of the signal is the same whether it is represented as a time-domain function or transformed to be

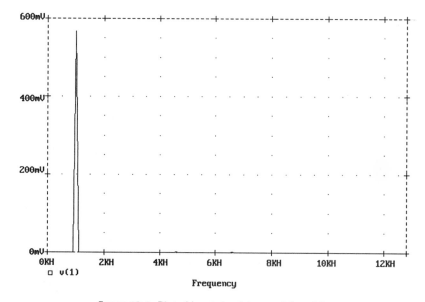

FIGURE 12-4 Plot of input signal (spectral domain).

a spectral-domain function (this is a result of Parseval's theorem). For example, if you want to know how much power there is between two frequencies, you integrate the square of the spectral amplitude between those two frequencies.

- The result from a discrete Fourier transform, such as the FFT, is **not the same** as the continuous Fourier transform values sampled at the same frequencies. Due to the lack of resolution of the DFT, each data point of the transform is actually the level of a "bin" (in frequency, for the transform of a time value) that extends halfway to the adjacent data points, with an area approximately equal to the value of the area under the Fourier transform for the same range of frequencies.

- Accordingly, the Fourier transform of a sine wave results in a delta function, that is, an infinitely high *spike* of zero width, at the frequency of the sine wave, whose area represents the amplitude of the sine wave. For the DFT (FFT) of the same sine wave, the spike has finite height and non-zero width, but the area is approximately the same. If more samples are used for the DFT, the transform has more resolution (it becomes more like the continuous Fourier transform) and the DFT spike gets taller and narrower, its area converging to the value of the delta function.

- For other sections of the spectrum, those not containing single frequency signals, you may as well consider the DFT (FFT) result to be the sampled value of the Fourier transform at that frequency.

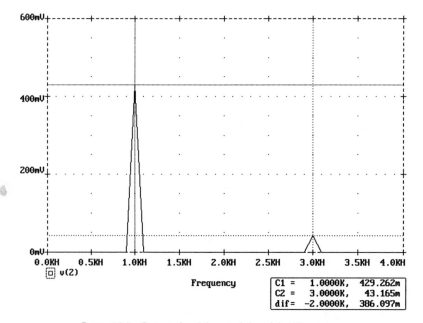

FIGURE 12-5 Output signal (spectral domain) with cursors.

These caveats are not unique to discrete Fourier transforms; similar precautions apply to the use of some types of spectrum analyzer equipment.

However, the waveforms resulting from computer simulations generally do not have continuous spectra. This is because the input waveforms are generally pure, such as a sinusoidal waveform, or have a definite set of harmonics, such as a pulse. There is no broadband source available in the simulator for transient response. And, since nonlinear circuits can convert only single frequency combinations into other single frequency combinations, the output waveforms will always contain only individual and distinct frequencies (like spectral lines, to use the analogy to light). Any response between these frequencies, as calculated by the FFT, is due to numerical noise in the interpolation and transform operations. So, we ignore the results of the FFT in these areas.

Now, since we are going to be looking at only the spikes in the transform results, there is no sense in manually calculating the area of these to find the signal power of each. The height of the spike is proportional to the number of samples going into the transform, so Probe scales the result for you. This means the height of the spike is really the calculated spectral amplitude, and this does **not** vary with the number of samples put into the FFT. For our example, the height of the spike is 567.55mV, which is within ½% of the amplitude we set for the sine wave in this example.

Looking at the transform of v(2), in Figure 12-5, we see the third harmonic "spectral line." Measuring the spike at 1KHz we get 429.262mV, which is, again, within ½% of the value we obtained using the harmonic decomposition provided by the .FOUR statement. For the spike at 3KHz we get 43.165mV, which is low by 7% from the value obtained using the harmonic decomposition. This error may be improved by using more samples.

The primary source of error in the transform comes from the interpolation done by Probe to obtain evenly spaced samples: that is, the time points in the transient run did not occur at the right places (this need not be true of transforms of a frequency analysis, where you can control the analysis interval). Using interpolated values is fine so long as you realize that this *smooths* the results, which is a way of saying that the signal is being processed, **before** being transformed, by a low-pass filter. This filter's transfer ratio is unity at DC, zero at the Nyquist critical frequency, and some other value in between. This is why the results we obtain for harmonic magnitude using the FFT in Probe are always lower than the results from the .FOUR statement. The effective bandpass of this smoothing filter is increased by using more samples in the FFT, which means forcing PSpice to take more time-steps.

EXERCISE 12.6-1

Improve the accuracy of the measurement of the third harmonic by doubling the number of steps in the Fourier transform. You will do this by decreasing the maximum step size of the transient simulation. Notice that the X-axis extends to twice the range as our example. How does the new spike at 3KHz compare with the value from the harmonic decomposition?

12.7 Intermodulation Distortion

Having just tried an example of calculating spectral values, let's broaden our scope to calculate intermodulation distortion. "Intermodulation" is the name for the process by which any signal processor, such as an amplifier, converts superimposed signals into signals having frequencies that are the sum and difference of the frequencies of the input signals. The nonlinearities that cause this to happen also produce distortion products at other frequencies, such as twice the frequency of signal A, minus the frequency of signal B, and similar combinations. Since most of these combinations are not harmonically related to the input signals, these distortion products are considered to be the most objectionable in, say, a high-fidelity audio system. There are many techniques for expressing the amount of distortion of this type; however, we shall look at a total calculation as a guide to other methods.

Using a variant of our previous example, we inject two signals into the polynomial function so that the peak input barely saturates the function. Using the frequencies of 800 Hertz and 1,000 Hertz to present one style of measuring intermodulation distortion, where the input frequencies are close together, the circuit file becomes

```
Intermodulation distortion
.opt itl5=0
i1   0 1 sin(0 .28 1000)
i2   0 1 sin(0 .28  800)
rin 1 0 1
e3   2 0 poly(1) (1,0) 0 1 0 -1
r3   2 0 1G
.tran 50u 50m 0 50u
.probe
.end
```

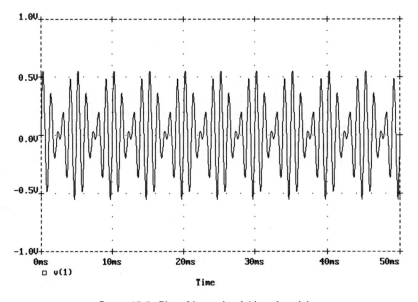

Figure 12-6 Plot of input signal (time domain).

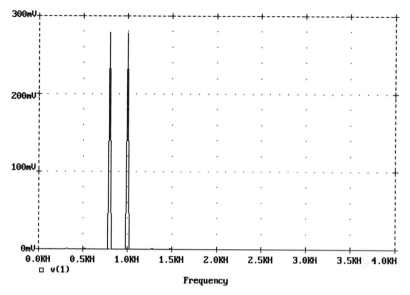

FIGURE 12-7 Plot of input signal (spectral domain).

After running PSpice, Probe is again used to transform the results into the signal spectra. We can display the input signal, as in Figure 12-6, to check that the simulation went well. Now, we transform the waveform to show the input spectra, as in Figure 12-7. At this time, you might want to try measuring the input spectrum, which is known. The peaks at 800 Hertz and 1,000 Hertz are single frequencies, so their height represents their energy content:

> 278.802mV at 800 Hertz

> 277.976mV at 1,000 Hertz

These values are very close to the 280mV input magnitude. Notice that the higher frequency signal's magnitude is lower, a manifestation of the smoothing filter resulting from using interpolated values. Now, looking at the output spectra, in Figure 12-8, we see many more spectral lines. Most of these are intermodulation distortion products. First, we measure the height of the fundamental outputs:

> 229.621mV at 800 Hertz

> 228.941mV at 1,000 Hertz

Then, there is harmonic distortion, most of which is at the third harmonic of the input:

> 5.2395mV at 2,400 Hertz

> 5.1015mV at 3,000 Hertz

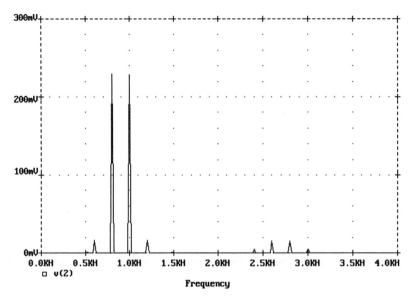

FIGURE 12-8 Plot of output signal (spectral domain).

These combine to yield harmonic distortion of

$$\sqrt{\frac{(5.2395\text{mV})^2 + (5.1015\text{mV})^2}{(229.621\text{mV})^2 + (228.941\text{mV})^2}} \approx 2.26\% \tag{12-2}$$

which looks good. Even though the signal reaches the clipping level, the individual distortion energies do not amount to much.

Finally, there is intermodulation distortion:

16.435mV at 600 Hertz

16.287mV at 1,200 Hertz

15.586mV at 2,600 Hertz

15.447mV at 2,800 Hertz

These combine in the same root-sum-square fashion to yield an intermodulation distortion of

$$\sqrt{\frac{(16.435\text{mV})^2 + (16.287\text{mV})^2 + (15.586\text{mV})^2 + (15.447\text{mV})^2}{(229.621\text{mV})^2 + (228.941\text{mV})^2}} \approx 9.83\% \tag{12-3}$$

which is over four times the combined distortion energy of the harmonic distortion.

Device models are SPICE's way of collecting the operating characteristics of circuit element or a *device*. So far we have worked only with fairly simple devices such as resistors and capacitors, and even these devices can make good use of having a model. But especially for active devices, such as diodes and transistors, it is essential to collect the numerous parameters describing how the device will behave and then refer to that particular set of parametric values by a shorthand name. This lets you label each instance of the device, in the circuit file, by a name that is convenient and mnemonic. Furthermore, when you decide to change the model parameters, this needs to be done only in the device model, and it will affect all of the devices in your circuit file that refer to that model.

13.1 THE .MODEL STATEMENT

The .MODEL statement sets aside a set of parametric values for reference by devices in PSpice. Not every device needs a model; for example, resistors not referring to a model are assumed to have a constant resistance value for all simulations. Every device that does refer to a model must have that model defined, which means it needs to have a .MODEL statement that completes the description for how the device will operate. The syntax for the statement is

.MODEL *name type_name* (⟦ *parameter_name* = *value* ...⟧)

The *name* is the shorthand label or "part name" by which you want to refer to the device. Often this is a manufacturer's part number, such as "MJE3055" for a transistor, or a descriptive name, such as "FILM" for a metal-film resistor. You may use any name that conforms to the naming conventions of the simulator; names must begin with a letter from the alphabet, and continue with alphabetic or numeric characters, or "_" and "$". For example, "2N3904" is usually modified to be "Q2N3904" to fit the naming conventions.

The *type_name* is a "device type" description, which may be one of the following for the linear devices:

CAP	capacitor
IND	inductor
RES	resistor

or one of the following for the semiconductor devices:

D	diode
NPN	NPN bipolar transistor
PNP	PNP bipolar transistor
LPNP	lateral PNP bipolar transistor
NJF	N-channel junction FET
PJF	P-channel junction FET
NMOS	N-channel MOSFET
PMOS	P-channel MOSFET
GASFET	N-channel GaAs MESFET

or one of the following for the "miscellaneous" device group:

CORE	nonlinear, magnetic core (transformer)
VSWITCH	voltage-controlled switch
ISWITCH	current-controlled switch

or one of the following for the digital interface and digital device group:

DINPUT	digital input device (receive from digital)
DOUTPUT	digital output device (transmit to digital)
UIO	digital I/O model
UGATE	standard gate
UTGATE	tri-state gate
UBTG	bidirectional transfer gate
UEFF	edge-triggered flip-flop
UGFF	gated flip-flop
UDLY	digital delay line
UPLD	programmable logic array
UROM	read-only memory
URAM	random access (read/write) memory
UADC	multi-bit analog-to-digital converter
UDAC	multi-bit digital-to-analog converter

Any of the parameters allowed for the device model are then defined. If you do not include a model parameter and value, there is a default value that will be used instead. Usually these default values are set to a convenient value that produces a typical operation, or are set so they have no effect on the device's operation (which means you may ignore them if they are of no interest). Let's look at what these parameters can do for your circuits.

13.2 MODELS FOR PASSIVE DEVICES

When SPICE users talk about *models*, they are usually referring to models for semiconductor devices and integrated functions. This is an important area; design groups using circuit simulators often maintain libraries of models for their work, and may even have some engineers whose only job is to develop new model sets. However, even simple devices may have models. For example, the model for the capacitor includes parameters for the following:

  c   which is the scaling factor (default value = 1)
  VC1 which is the linear voltage coefficient (default value = 0)
  VC2 which is the quadratic voltage coefficient (default value = 0)

and others, which we will look at later. Already you may suspect that we will be able to define a type of capacitor whose capacitance will vary with the voltage impressed on the plates by setting the voltage coefficients.

To see how to use the device models, let's recall the syntax for including a capacitor in a circuit file:

  Cname node node [[*model_name*]] *value*

This is different from the way you are used to writing it, since it includes the *model_name*, which is an optional item that we have not used before. So, you can have a circuit with capacitors, some of which may include a reference to a model as in this fragment of a circuit file:

```
C5 2 7 .015
C6 3 5 cx 2
C7 4 6 cx 1
.model cx cap(vc1=.1 c=.001)
```

In this case, the value of C5 is 0.015 farad but the values for C6 and C7 depend on the model parameters. Their values are calculated by the formula

$$capacitance = value \cdot c \cdot \left(1 + VC1 \cdot voltage + VC2 \cdot voltage^2\right)$$ (13-1)

This means that, with no voltage across the capacitors, the value for C6 is 0.002 farad and the value for C7 is 0.001 farad. As the voltage varies, C6 and C7 will change their values by 10%/volt, where the voltage is the difference between the first node and the second node (so it matters, now, which way you connect the capacitor).

13.3 SCALING COMPONENT VALUES

Using the previous example circuit fragment, since both C6 and C7 refer to the cx capacitor model, you may scale their values relative to each other; that is, regardless of the value calculated by the formula above, the value of C6 will **always** be twice the value of C7 (with zero bias, since this example shows a capacitor model that is also voltage dependent). You may shift the values of a whole set of capacitors in your circuit without changing their relative values, by changing the model to, say

```
.model cx cap(vc1=.1 c=.002)
```

to double their zero-bias values to 0.004 farad and 0.002 farad, respectively.

Another way to set up capacitors that scale is to start with the actual component values you would normally use; for example:

```
C5 2 7 .015
C6 3 5 cx .002
C7 4 6 cx .001
.model cx cap(c=1)
```

so that the values for C6 and C7 are still 0.002 and 0.001, respectively, because the multiplier parameter is now 1. Then, should you decide later to increase the values of the cx capacitors by 10%, you would change the model to

```
.model cx cap(c=1.1)
```

You may use as many models as you like in a circuit file, although each component may refer to only one model. If you are crafty, you may find ways to set up "generic" circuit blocks, such as filter modules, which are customized by setting the value of the multiplier parameter.

EXERCISE 13.3-1

Write a circuit fragment that duplicates the features of the example in this section, but uses a .PARAM statement and an {*expression*} to set each capacitor's value. This is the preferred way to implement component values that are related by a formula, as complex situations can be handled. See §2.4 (page 10) to refresh your memory.

As you might have suspected, the resistors and inductors in PSpice may have models also. Just like the C multiplier:

R is the resistance multiplier parameter in the RES model, and

L is the inductance multiplier parameter in the IND model.

This means we could have used scaling models for all of the elements in our LC-filter example, as shown (for one of the filter sections):

```
*   Q = 1
R2 1 4 rmod 1
L2 4 5 lmod 1
C2 5 0 cmod 1
.model rmod res(r=100)
.model lmod ind(l=10m)
.model cmod cap(c=1u)
```

If we did this correctly for all of the filter sections, we could shift the natural frequency of the filters while keeping the same Q values.

13.4 SWEEPING COMPONENT VALUES

In PSpice, the .DC statement has been generalized to sweep model parameters that in turn will sweep the component value, as well as the SPICE function of sweeping voltages or currents. This is accomplished by extending the meaning of *source_name* to include references to model parameters. For example:

```
.DC vin 2 12 2
```

sweeps the value of the voltage source vin from 2 volts to 12 volts, in 2-volt increments. By example, we could also sweep the resistance multiplier in rmod with the following circuit file fragment:

```
R7 4 6 rmod 1
.model rmod res(r=100)
.DC res rmod(r) 100 150 10
```

which will sweep the R parameter of the RES model named RMOD, starting with R = 100, in increments of 10, until R = 150. This changes the multiplier for R7 and any other resistors that reference this model. See the .DC statement description in Appendix B (page 218) for more details on sweeping the values of model parameters.

EXERCISE 13.4-1

Since you have just refreshed your memory on the details of the .DC statement, write a circuit fragment that duplicates the features of the example from this section, but uses a .PARAM statement to define a control parameter, a .DC statement to sweep the parameter's value, and an {*expression*} to set the resistor's value. This is the preferred way to sweep component values, as complex situations can be handled.

13.5 TEMPERATURE ANALYSIS

Another common way to check a circuit is to operate it at different temperatures to verify that certain performance standards are met. SPICE and PSpice both include a control statement of the form

.TEMP *value...*

with a list of values specifying the temperatures, in degrees Centigrade, at which all of the other analyses (AC, DC, transient) are to be run; that is, the .TEMP statement acts as an "outer loop" for all of the other analyses. When the temperature is changed, PSpice recalculates internal values using the new temperature (for example, the noise contribution for resistors involves the factor $4 \cdot k \cdot T$) and makes adjustments where the device models have a temperature dependence. For example, the device models for the capacitor, inductor, and resistor all have parameters for temperature dependence, which are

TC1 a linear dependence on the change in temperature, in %/°C

TC2 a quadratic dependence on the change in temperature, in %/°C^2

so that, just looking at the temperature-related factors, the formula for the value of a resistor with a model is

$$\text{resistance} = value \cdot \text{R} \cdot \left(1 + \text{TC1} \cdot \Delta\text{T} + \text{TC2} \cdot (\Delta\text{T})^2\right) \quad\quad (13\text{-}2)$$

where ΔT = .TEMP_*value* − Tnom (Tnom is nominally 27°C and may be set using the .OPTIONS statement, though this is rarely done). The capacitor and inductor follow the same form. Other devices, such as transistors, have temperature dependencies built into their more complicated physical models, so that the user does not have to include any factors directly for operation at different temperatures.

13.6 SWEEPING TEMPERATURE

Similar to the way PSpice sweeps component values, you may also command PSpice to sweep the value for temperature. The calculations performed are the same as those for the .TEMP statement; however, the simulation is done as a DC sweep so that graphical results are available to Probe. Moreover, since you are allowed to nest one DC sweep within another, you will be able to sweep a source or component value while stepping the temperature (or vice versa)!

The form of the .DC statement to sweep temperature is similar to what we have used for sweeping component values. For example:

```
.DC TEMP 30 50 5
```

will sweep temperature, called TEMP by PSpice, from 30°C to 50°C in 5°C increments.

CHAPTER 14

MONTE CARLO SIMULATION

Congratulations! Your circuit design is finished and you are satisfied with its predicted (simulated) performance. And yet, you are concerned: How well will this design perform when built with real components, which will not have exactly the values (parameters) you used in your simulations? After all, the point of circuit design is not great simulations but to build great circuits, or at least to build circuits that will operate as specified and be useful.

"Classic" circuit simulation has two limitations that keep you from understanding the effect of component variations:

- You will usually change just one component value and re-simulate. Exploring the entire range of possible changes in performance is virtually impossible, even for a modest circuit. You cannot realistically determine the risk of your circuit design being unacceptable.

- A "what if" analysis only provides isolated predictions without indicating the likelihood of achieving any particular result. That is, a "what if" simulation will tell you what is **possible** but not what is **probable**.

PSpice overcomes both of these limitations, and can take the guesswork out of using circuit simulation to provide insight into production realities, by combining *Monte Carlo methods*[†] with circuit simulation. This is combination is commonly called *Monte Carlo simulation*.

To understand how a simple Monte Carlo method works we can use an easy geometry example: Calculating the area bounded by a curve, for example an ellipse

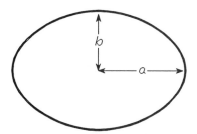

FIGURE 14-1   An ellipse.

[†] The phrase "Monte Carlo methods" was coined in the late 1940s by mathematicians Stanislaw Ulam and Nicholas Metropolis to convey the nature of the techniques they were developing. Referencing the gaming tables of Monaco provided the desired contrast with algebraic approaches.

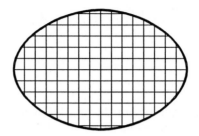

FIGURE 14-2 Subdividing to calculate area.

with semi-axes *a* and *b* as shown in Figure 14-1. A primitive approach would tile the interior of the curve, filling it with small squares and totaling the area to get an approximate answer, as shown in Figure 14-2. Or we could use integral calculus to make the tiling procedure exact, in which case we find that the area is πab.

Rather than tiling the interior of the ellipse with shapes of known area, suppose we compare the unknown interior to a known exterior region: A square with sides 2*a* will contain the ellipse (see Figure 14-3) and all we need, now, is an independent method for determining the ratio of the two areas to calculate the area of the ellipse. This is the idea behind integral Monte Carlo methods that, in simple form, can approximate the ratios of areas by randomly scattering points inside the reference area, as in Figure 14-3. If c_0 of the total *c* points fall within the curve, then the ratio we want is approximated by c_0/c, so the interior area is approximated by $4a^2c_0/c$.

This technique provides a way to estimate any area, or volume, etc., by creating a "game" to yield an answer that **increases in accuracy the longer the game is played** (that is, as the total number of sample points increases). In fact, the statistical error of the answer in our example is proportional to $1/c^{1/2}$. And, so long as we can determine which sample points lie inside, the curve can be complex and this Monte Carlo method will provide a useful answer even when analytic techniques fail and give no answer at all.

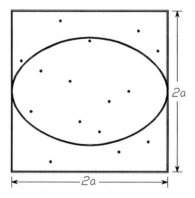

FIGURE 14-3 Statistical method for calculating area.

While the purely random approach to area approximation is inefficient compared to other Monte Carlo methods, it demonstrates how a statistical technique yields a useful answer... indeed an answer that may not be possible analytically.

14.1 PROBABILITY DISTRIBUTION

Suppose someone accidentally spilled a large bucket of sand on the floor, and you wanted to describe this pile of sand. You would be likely to mention three things: its general shape, which may be like a rounded mound or a sharp cone; its location, say relative to the door or a wall; its dispersion, which for a cone shape may be confined or for a flat shape may cover much of the floor. This way of describing a pile of sand is similar to describing the distribution of measured quantities, like test scores and component values. Shape information has certain names, like *normal*, *symmetrical*, *negatively skewed*, and *bimodal*. Location is described just as the pile of sand was by using a reference and a measurement to the center, or central tendency, of the shape. Dispersion, or variabililty, is measured from this center may record the average distance of items (grains of sand) from the center.

It might be easier to understand how a distribution describes probability by considering an example: Say you want to look at the distribution of actual values of resistors marked 270 ohms, using a small, analog meter. You start with 18 resistors, and sort them by measured value. You work quickly, so you only read the meter to the nearest 10 ohms. Then you chart the frequency distribution of the measurements by listing the measured values along the horizontal axis, and the count (or frequency) of resistors for each measured value on the vertical axis, as shown in Figure 14-4.

You could also chart these measurements as a probability distribution by scaling the vertical axis by the total number of resistors (dividing each count by 18). This creates the chart seen in Figure 14-5.

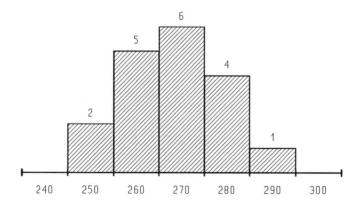

FIGURE 14-4   Resistance measured to nearest 10 ohms.

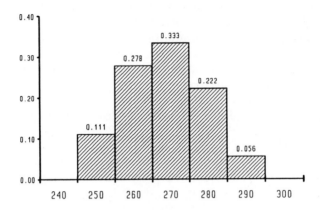

FIGURE 14-5   Probability distribution to nearest 10 ohms.

Since we measured a small number of resistors, and only to the nearest 10 ohms, we can imagine that if we had time to measure more resistors to the nearest ohm, we could construct a "smoother" probability distribution for these resistors, as shown in Figure 14-6.

Following this idea, we could imagine measuring an infinite number of resistors, precisely, to obtain a smooth curve representing the probability distribution of all 270 ohm resistors (also Figure 14-6, the smooth "bell" curve). It is tempting to say that a probability distribution, say $f(x)$, represents the probability of observing the measured value x, but this is not really true since the measured value x is a continuous variable: The probability of measuring exactly x is zero. But it is proper to say that the probability of measuring a value in any region, say, between x and $x+dx$, is the integral of $f(x)$ (the area under the curve) between x and $x+dx$. If you look back at the previous figures showing the measurements made on the resistors, the charted data (a histogram) clearly showed this interpretation.

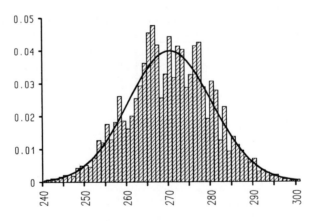

FIGURE 14-6   Probability distributions.

However, the resistance value of a resistor is the combination of many variables when it was manufactured. Let's say these were carbon-composition resistors, in which case the material that makes up the "carbon-composition" will vary in conductivity. The change in conductivity, from resistor to resistor, has a probability distribution. The geometry of the "carbon-composition" cylinder, inside the protective coating, will also vary, slightly, to affect the overall resistance of the device. This variation in size has a different probability distribution from the change in conductivity of the material. Also, how well the leads are connected to the "carbon-composition" is a source of more resistance. This variation in contact resistance is yet another, independent, probability distribution to consider. You may be able to dissect these three sources of variation into more basic causes for "why resistors vary."

When the resistors are made, though, the final probability distribution is the one we measured, and it is a combination of all the separate distributions. You could mathematically construct the overall distribution, assuming you had formulas for the sub-distributions, but as the system you are considering gets larger the math quickly becomes overwhelming. This is where statistical techniques, such as a Monte Carlo method, become useful in creating a practical answer about the probability distribution of a complicated system.

14.2 THE MONTE CARLO SIMULATION TECHNIQUE

When integral Monte Carlo methods are applied to problems like circuit simulation, the purely random selection of "what if" values for parameters in the circuit is inefficient. Instead, an importance-sampling procedure is employed to use knowledge of the probability distribution to, in effect, focus calculation effort on the "important" ranges of the parameter. The random value generator is controlled by the probability distribution of the parameter instead using an unbiased distribution over the entire range of possible values. In other words, the computer generated values are random but, as a set, are statistically the same as actual measurements. This is done for every varying parameter the user wants to specify. Meanwhile, a series of simulations is performed, with all the values varying accordingly, and the outputs of interest are inspected. And here is the important fact: The mathematics supporting this Monte Carlo method proves that the probability distribution of the simulated results will be statistically the same as actual measurements of a real circuit. Monte Carlo simulation is simply a statistical technique that turns "what if" into "what's likely" by an intelligent use of the "what if" approach.

But, as we saw earlier with measuring resistors, there is a trade-off between effort and confidence:

- you can assume all the resistors are 270 ohms,

- you can measure a small quantity of them to the nearest 10 ohms,

- you can make more measurements to the nearest ohm, or

- you can assume a smooth representation.

And so it is with the Monte Carlo methods; simulation "effort" is used to gain confidence in a practical result that cannot be modeled exactly. From this effort you get two rewards: You can see what range of outputs is realistically possible, and you can gain confidence in the probability that "good" outputs (or the risk that "bad" outputs) will occur for a real circuit. In fact, your confidence increases in proportion to the square-root of simulation effort. Fortunately, the computer does all the work... all you have to do is wait.

We believe that it is especially useful to have a computer do this kind of work for the risk of unacceptable circuit operation is generally the product of the independent probabilities of many components in the circuit, and the actual components involved are likely to change from case to case of unacceptable operation. This is a difficult thing for the human brain to judge. Generally we consider only the worst possible cases, wherein all the variables are against us, which usually has the least risk of happening. This fact of probabilities works to the advantage of railroads, for example, were if the probability that the railroad engineer is asleep is one small number, and the probability that the automatic system malfunctions is another small number, then the probability of a wreck (asleep and a malfunction) is the product of these two small numbers, which is a much smaller number!

14.3 COMMONLY-USED PROBABILITY DISTRIBUTIONS

Selecting a probability distribution can be difficult for the novice, since your engineering instincts suggest that you be as exacting as possible. However, since we are usually trying to gain insight, rather than an exact answer, we are way ahead of the game just having Monte Carlo simulation available; selecting the "best" probability distribution is a "second-order effect" for most purposes. The easiest way to be comfortable with your choice is to have historical data, such as a plot, of measured values for the item that varies. Or, you can list everything you know about the conditions that make the variable a variable (and not a constant) and use this list to judge which probability distribution would be appropriate.

The most commonly-used probability distributions have names you may remember from mathematics class:

- *Uniform* distributions have all values, between a minimum and maximum value, occurring with equal likelihood. This is the most commonly-used probability distribution because so little needs to be known about the variable. Also, this distribution may be an appropriate choice for electronic devices that have been sorted by value, before being marked, so that you know the device values are within a certain range. PSpice is preprogrammed to provide this distribution using the keyword UNIFORM.

- *Normal* (or *Gaussian*) distributions are useful for describing the variability in natural phenomena, where an unlimited number of underlying processes may be at work to create the measured variable, such as population growth or the inflation rate. This probability distribution is particularly useful since it represents the natural result of large, random systems: Regardless of the probability distributions of the underlying processes, as you combine an unlimited number of them the result tends toward a normal distribution. So this distribution describes the theoretical tendency of any system, which makes it a reasonable choice for describing most variables (unless you know about particular constraints, like presorting that eliminates variations beyond a certain amount). PSpice is preprogrammed to provide this distribution, using the keyword GAUSS, over a range of ±3 standard deviations about the mean value, which accounts for 99.72% of the full distribution.

- *Triangular* distributions describe situations where you know the minimum, maximum, and most likely values, and the values near the minimum and maximum are less likely to occur.

14.4 LESS COMMONLY-USED PROBABILITY DISTRIBUTIONS

Scanning other texts on probability distributions, you find many other types. Some of these are useful for circuit simulation purposes, but most are not. In particular, you will find distributions that describe the outcome of trials (or experiments) and occurrence of events. These are discrete probability distributions:

- *Binomial* distributions describe the number of times a particular event occurs in a fixed number of trials (for example, how many "heads" in a fixed number of coin tosses or how many defective units in a fixed production run).

- *Poisson* distributions describe the number of times an event occurs in a fixed interval (for example, the number of customer complaints per day or the number of defects per square meter of material).

- *Geometric* distributions describe the number of trials until the first desired outcome occurs (for example, the number of spins of a roulette wheel needed to win or the number of oil wells drilled to get a producing well).

- *Hypergeometric* distributions are similar to binomial distributions except the trials are not independent, and are called "trials without replacement" (for example, how many defective units in a fixed production run remain after some of the units have been shipped).

Other probability distributions are useful to circuit simulation, however their derivation and use may prove beyond the needs of getting insight into your circuit:

- *Lognormal* distributions describe positively skewed probability distributions, such as real estate values (that is, values that cannot go below a minimum value, such as zero, but may increase without limit).

- *Exponential* distributions describe events that occur at random in time (for example, the time between component failures or customer complaints). This distribution is only for memoryless situations, in which any event does not influence future events.

- *Weilbull* (or *Rayleigh*) distributions describe the results of reliability studies and quality control tests. When the Weilbull "shape" parameter equals two the distribution is called a Rayleigh distribution, which happens to be the root-mean-square (RMS) addition of two independent normal distributions (each having a zero mean and the same standard deviation). Rayleigh distributions are particularly useful for describing communications problems like signal fade due to multipath reflections.

- *Beta* distributions are generally used to "curve fit" to empirical data that ranges between zero and a positive value. Three parameters provide a useful combination of scale and skew to fit many situations.

- *Gamma* (also *Erlang* and *Chi-square*) distributions are commonly used to describe the duration until the *n*th occurrence of a Poisson process.

- *Logistic* distributions are used to describe the evolution of a process (for example, growth of a population over time or the rate of chemical reactions).

- *Pareto* distributions are used to describe empirical phenomena such as stock price fluctuations and errors in communication channels.

- *Extreme Value* (or *Gumbel*) distributions describe the largest value of response over a period of time (for example, rainfall, flooding, and earthquakes).

14.5 DEFINING PARAMETER ASSUMPTIONS

In PSpice's implementation of Monte Carlo simulation, all parameter variations are defined using device models. This means each device that you want to vary, even a resistor, will have a model.

Now it may have occurred to you that some devices share model information, which means that you have several of the same types of device in your circuit. This is fine, but it may not have occurred to you that sometimes the variations in similar devices track each other (or "are correlated"). This can happen if the devices are manufactured at the same time, as happens with individual devices in an integrated circuit, transistor array, or a thin-film device. However, even the devices in an integrated circuit will show a small variation with respect to each other.

To handle these situations, two types of parameter variation are allowed: (individual) device variations and (as a group) lot variations. Either type of variation, or both, may be specified when describing how parameters vary. For example, you would use device variations to describe how discrete carbon-composition resistors are statistically modeled, but you would use lot variations to describe the dominant variability in the elements of the thin-film R-2R ladder network in a digital-to-analog converter.

PSpice has two predefined probability distributions: the uniform distribution and the normal (or Gaussian) distribution. Any other distribution you may want needs to be defined, along with a unique name, using the .DISTRIBUTION statement. This statement allows up to 100 number pairs, representing variation and probability, which are used to define the corners of a piecewise linear description of the probability distribution you want. If you know the probability distribution as a formula, you will need to develop a suitable approximation using a piecewise linear curve. The variation values are allowed over the range from −1 to +1, and the area under the curve is normalized (internal to PSpice) so that the probability values may represent whatever is convenient. The variation range is individually scaled for each parameter using it. (The .DISTRIBUTION statement is described in Appendix B.)

At first you might think that requiring a piecewise linear description of the probability distribution is cumbersome since you have learned about the smooth, approximating curves typically used to describe probability distributions. However, the corners of the piecewise linear description may simply be the collected measurements of actual devices — as we saw earlier in this chapter. Using this data directly is a useful approximation in itself, and removes any judgments that would have to be made about the "goodness" of a smooth curve as an approximation.

For example, we may want to create a probability distribution for the earlier example involving 270 ohm resistors measured to the nearest 10 ohms. From the spread of the measurements we decide that these were ±10% devices, so when creating the piecewise linear approximation we will want to map ±10% onto −1 to +1 in the description: −10% will map to −1 and +10% will map to +1. Every 10 ohms will correspond to roughly $10/270 \cdot 10\% = 0.37$ of variation. So, our new probability definition will look like:

```
.DISTRIBUTION my_data (-1,0) (-.74,2) (-.37,5) (0,6)
+   (.37,4) (.74,1) (1,0)
```

The .MODEL statement that uses this probability distribution may look like:

```
.MODEL my_resistor RES(R=1 DEV/my_data 10%)
```

And the circuit device that uses this model may look like:

```
R17 gnd in my_resistor 270
```

The model parameter format is somewhat complicated by the range of options you might need to describe complex situations. Each model parameter in a .MODEL statement has the form:

name = value [[*tolerance_spec*]]

and it is the [[*tolerance_spec*]] that can be complicated. For this you will usually use a form like:

{DEV|LOT} *value*[[%]]

and use the default distribution, which is a uniform distribution (unless changed by the .OPTIONS statement). Including a percent sign "%" means the value is relative, as a percentage, and not an absolute value (in the same units as the parameter itself). The entire [[*tolerance_spec*]] has the following format:

[[DEV [[*track_&_dist*]] *value* [[%]]]]   [[LOT [[*track_&_dist*]] *value* [[%]]]]

which has identical sections for specifying device (DEV) and lot (LOT) deviation information. As mentioned before, the device deviations are independent and the lot deviations track between devices that reference the same .MODEL statement. But remember, this stuff is for Monte Carlo, not regular, simulations.

The optional [[*track_&_dist*]] information follows the keywords DEV or LOT with a slash "/" and **no spaces**, using the following format:

[[/*gen#*]]   [[/*distribution*]]

The *distribution* is simply a name, either predefined (UNIFORM and GAUSS) or one you have defined with a .DISTRIBUTION statement. The *gen#* is a digit, 0 through 9, that selects one of ten random number generators to let you synchronize deviations between model parameters. You might think of this as defining a kind of lot tolerance between parameters in the same model, or different models, just like the regular lot tolerance between devices using the same model. By definition, there are only twenty of these special random number generators available to your circuit; ten for all DEV specifications and ten for all LOT specifications. Of course, if you do not use *gen#*, then the deviation is set by an independent random number.

As an example of why you might correlate deviations, let's say you were modeling an R-2R ladder with variability in the resistive material as well as variabililty in the contact resistance (where the resistors connect to conductors). The ladder is built using an integrated process, such as thick-film, thin-film, or an IC technique. Instead of using separate resistors in the circuit description to model the contact resistance, we can fold all the modeling into correlated models. The circuit description for the ladder elements might look like:

```
RA 25 26 R2mod 1  ; R2 leg
RB 26 27 R1mod 1  ; R1 leg
```

and so on. Note that the resistance specification has been set to 1, which will be multiplied by the model's R parameter to obtain the actual resistance. Now, the models will look like:

```
.MODEL R1mod res (R=10K DEV 1 LOT/3 5%)  ; note DEV = 1ohm
.MODEL R2mod res (R=20K DEV 1 LOT/3 5%)  ; note LOT = 5%
```

so that the models set the nominal resistance level, and an individual deviation of ±1 ohm for the contact resistance (assuming the default distribution type is UNIFORM). The lot deviation is ±5% for changes in the material that makes up the bulk of the device and the two models are synchronized in modeling their lot deviations, using the number "3" generator, which is as it should be for an integrated process. Synchronized deviations might prove useful for MOS circuit simulation, for example, where every MOS model's oxide thickness should be correlated.

14.6 RUNNING A MONTE CARLO SIMULATION

The .MC statement causes Monte Carlo simulation to be done, in due course, with other work by PSpice. There are four items PSpice needs:

- how many iterations, or runs, will be made,

- which analysis will be done, and you'll have a statement in the file describing that analysis,

- how to organize information coming out of the runs, and

- the kind and amount of output generated.

There is also an option for setting the start-up of the random number generator. The .MC statement has the following form:

.MC *#runs* {AC|DC|TRAN} *output_value function* [[*option...*]] [[SEED=*value*]]

The *output_value* has the same specification as in the .PRINT or .PLOT statements. This output is "filtered" by *function*, which condenses the output from each run, an entire waveform, into just one number! That number is used to organize all the printed output, from Monte Carlo simulation — a process called "collating" in the data processing field. Sometimes the function that organizes results is referred to as a "collating function" because it enables the process of sorting the curves generated by PSpice. The available collating functions are:

- YMAX, which finds the greatest difference (plus or minus) along the output curve with respect to the nominal run's output.

- MAX, which finds the maximum (most positive) value along the output curve.

- MIN, which finds the minimum (most negative) value along the output curve.

- RISE_EDGE(*value*), which finds the first positive-going crossing through the threshold *value* along the output curve. Preceding the threshold there must be at least one analysis point below *value* and the reported output is the analysis point where the curve rises above *value*.

- FALL_EDGE(*value*), which finds the first negative-going crossing through the threshold *value* along the output curve. Preceding the threshold there must be at least one analysis point above *value* and the reported output is the analysis point where the curve falls below *value*.

The [[*option...*]] specification describes what kind of output will be generated, in the output file and for Probe, and limits the range over which the collating function operates. The [[SEED=*value*]] changes sequence generated by the random number generator. These options are described in Appendix B, and the normal option used with Monte Carlo analysis is OUTPUT(ALL) to save all of PSpice's output. Be sure to include an OUTPUT(*type*) option to save **any** output.

The typical Monte Carlo analysis will be run with a statement like:

```
.MC 50 AC VM(8) YMAX OUTPUT(ALL)
```

and there will be a .PROBE statement in the file to collect output for Probe to display. This statement specifies that 50 small-signal simulations will be attempted (one nominal and 49 for Monte Carlo), that the output VM(8) will be used for collating the run information in the output file (with YMAX as the collating function), and that all the simulation outputs will be available for any .PRINT, .PLOT and .PROBE commands in the circuit file.

14.7 Interpreting the Results

Setting the options in the .MC statement can fill the output file with pages of information detailing parameter values used for each run and how the runs compared to each other. This output will be interesting for when you need to discover what set of parameter values gave a particular run's output. Mostly, though, you will be looking at curves and histograms in Probe.

Probe has its own version of collating functions, with several useful ones already defined, as part of Probe's "performance analysis" feature. Probe has a meta-language (called "goal functions") for moving across a curve, identifying and saving interesting values or segments, and performing a calculation to arrive at a single number to represent a significant feature of the waveform. The predefined goal functions include: Max, Min, CenterFreq, GainMargin, PhaseMargin, Risetime, Falltime, Period, PulseWidth, Overshoot, and other functions. Goal functions, by themselves, are handy for extracting numbers from waveforms and automating data reduction. Probe's "performance analysis" will display traces of extracted value versus a swept parameter, thus showing a curve of some performance of interest to the user. However, since Monte Carlo simulation is a random collection with no swept parameter, Probe's "performance analysis" presents results as histograms.

To create histograms, we need to derive a single value from each Monte Carlo run. Probe will then display the collected values by sorting them into bins of values that are close to each other. The number of bins can be set using a Probe menu. The number of data points in each bin is displayed by the height of the bar on the chart Probe displays. Typically, you will want the percentage of all runs that yielded a particular result, which is the way Probe displays histograms.

Having selected the data from all the Monte Carlo runs, the menu commands that lead you to a histogram are: X_axis, Performance_analysis, Add_trace.

At this point, you may select function key F4 to find what circuit values and goal functions are available to you. Having entered a formula, as we will see in the example (later), Probe calculates the histograms and displays something like Figure 14-7.

Below the chart is a box that displays some common statistical metrics for your histogram:

- **n samples** is the number of extracted data values, which will be from all the Monte Carlo runs, including the nominal run (if you have selected to load all runs when starting Probe).

- **n divisions** is the number of bins in the histogram. This can be changed by selecting menu items X_axis, Options, Histogram_divisions. Note: The divisions span the range of the extracted data values and have equal width. Also, empty bins do not show in the display.

- **mean** is the arithmetic mean of the extracted data values (sum of these values divided by the number of values). It is the most stable statistical measure of the central tendency of a set of samples.

- **sigma** is the *standard deviation* or the root-mean-square (RMS) difference of the extracted data values from their mean value. Its units are the same as the data and it measures the variability of these values. If your histogram appears to be approximating a normal (Gaussian) distribution, then you can quickly calculate what percentage of circuits are likely to fall in any given range of operation. Also, it can be shown that, statistically,

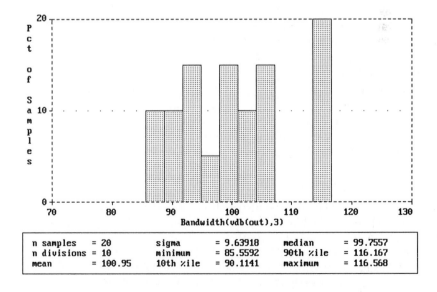

n samples	= 20	sigma	= 9.63918	median	= 99.7557
n divisions	= 10	minimum	= 85.5592	90th %ile	= 116.167
mean	= 100.95	10th %ile	= 90.1141	maximum	= 116.568

FIGURE 14-7 Typical histogram display using Probe.

the error in the calculated mean is not likely to be greater than the standard deviation divided by the square-root of the number of samples.

*   **minimum** is the smallest data value extracted.

*   **10th %ile** (the "tenth percentile") describes the portion of the distribution below which 10% of the extracted data values are placed, and above which 90% of the extracted data values are placed.

*   **median** describes the value that divides the distribution evenly with an equal number of data values above and below the median. It is also the "fiftieth percentile" of the distribution.

*   **90th %ile** (the "ninetieth percentile") describes the portion of the distribution below which 90% of the extracted data values are placed, and above which 10% of the extracted data values are placed.

*   **maximum** is the largest data value extracted.

Having these measures, and a histogram, is more helpful than having a list of frequency bandwidths for these runs.

14.8 Other Statistical Metrics

Probe provides the main statistical metrics with its "performance analysis" display of histograms. There are other statistical measures that might prove useful for comparing circuit operation, as well as other jargon that might be unfamiliar:

*   *Mode* is the value at which a distribution achieves its maximum value. Some distributions may have another local maximum, in which case the distribution is called "bimodal."

*   *Variance* is the second moment about the mean, which equals the square of the standard deviation. Obviously, variance and standard deviation both measure the same thing but standard deviation is more useful since its units are the same as the distribution's data.

*   *Coefficient of variability* refers to variability relative to the mean value, and is measured by the ratio of standard deviation to the mean.

*   *Coefficient of kurtosis* refers to the peak-edness of a distribution, and is measured by the ratio of the forth moment about the mean to the variance squared. The normal distribution has a kurtosis coefficient of 3, so values greater than 3 are *leptokurtic* (meaning peaked) and values less than 3 are *platykurtic* (meaning flat).

*   *Coefficient of skewness* refers to un-symmetrical nature of a distribution, and is measured by the ratio of the third moment about the mean to the standard deviation cubed. The normal distribution, being symmetrical, has

a skewness coefficient of zero. Distributions with positive coefficients have their mode less than their mean, and are called "positively skewed." A skewness coefficient greater than ±1 indicates highly skewed distribution, while a value between ±0.5 indicates a fairly symmetrical distribution.

14.9 AN EXAMPLE: TWO FILTERS

Since we are "getting our feet wet" using Monte Carlo simulation, let's choose a really simple circuit example and see what insight the technique will provide. A simple tank circuit (RLC resonator) will do the job, and we will simulate the variations in center frequency that might occur in manufacturing the circuit. Then, for contrast, we will replace the inductor of the circuit with a gyrator (an active circuit with the same impedance characteristics as an inductor) and see if a non-inductor implementation of the circuit is an improvement.

The circuit, shown in Figure 14-8, has a simple circuit file:

```
* RLC example
V1 in 0 AC 1
R1 in 1 1.5915494K
C1 1   0 1.5915494u
L1 1   0 15.915494m
.ac LIN 401 800Hz 1.2KHz
.probe V(1)
.end
```

Note that all the devices' values are $\pi/2$ time some power of ten, to set the nominal center frequency at 1KHz and have the Q equal to ten. This makes the Probe plots simpler to look at for understanding the results of Monte Carlo (after all, we're not learning about RLC circuits). Also, the small-signal analysis is done at 1Hz intervals so Probe can determine the center frequency to within 1Hz. This will be close enough for this example.

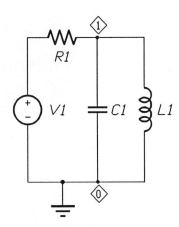

FIGURE 14-8   RLC filter circuit.

When we add the device models to specify component tolerances, our circuit file changes to:

```
* RLC example ready for Monte Carlo
V1 in 0 AC 1
R1 in 1 RMOD 1.5915494K
C1 1  0 CMOD 1.5915494u
L1 1  0 LMOD 15.915494m
.model RMOD res(R=1 dev/uniform 10%)
.model CMOD cap(C=1 dev/uniform 10%)
.model LMOD ind(L=1 dev/uniform 10%)
.ac LIN 401 800Hz 1.2KHz
.mc 200 AC V(1) YMAX output(ALL)
.probe V(1)
.end
```

Now each passive component has its own model, and all the components have been given a ±10% uniform variation in their resistance, capacitance, or inductance, respectively. Also, a .MC statement has been added to tell PSpice to run the small-signal simulation 200 times. The first run uses the nominal values and the other 199 runs use component variations specified by the tolerance information in the models and the importance sampling algorithm employed by the Monte Carlo technique. The collating function, YMAX, specifies what sorting will be done for the printed output file, but we will be ignoring this output. Finally, the voltage from one node will be collected for Probe.

Running Probe, we select all the AC runs and the menu commands X_axis and Add_trace to display all traces of output V(1) from the Monte Carlo runs. As we can see from Figure 14-9, this display shows a "smear" of traces. We can see the

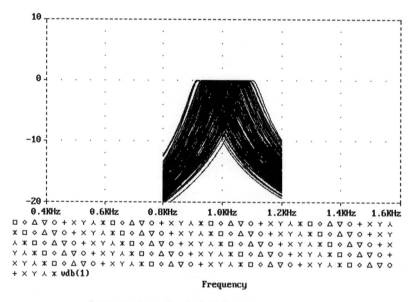

FIGURE 14-9 Output displayed as frequency response.

rough envelope of operation, but we get no sense of the density of the traces or whether there is any tendency for the circuit to operate near its nominal design.

Now we clear the traces and select the menu commands X_axis, Performance_analysis, and Add_trace to get us to the point of creating a histogram. At this point, you may select function key F4 to find what circuit values and goal functions are available to you. With goal functions enabled, we will use the CenterFreq function to search for the center frequency of each trace and save that value. For each trace, the CenterFreq function looks around the peak for a certain roll-off (an amount you specify) and then averages those two frequencies to derive the center frequency. For most circuits, this approach is adequate.

So, specifying:

    CenterFreq(VDB(1),3)

will find the 3-db points of the frequency response and average these to get the center frequency for that trace. All the traces are processed and the "answers" are plotted using a histogram to chart how many center frequencies fall in certain ranges. This is shown in Figure 14-10.

The first thing you might notice is that the distribution of center frequencies is greater toward the design frequency of 1KHz, even though the component deviations were uniform across a ±10% range. This happens because it is less likely for the components to achieve their maximum deviation at the same time (that is, for any given simulation). In fact, this notion is contained within a mathematical statistics theorem called the *central limit theorem*. The central limit theorem says (loosely) that as you include more components with deviations to your circuit, regardless of how

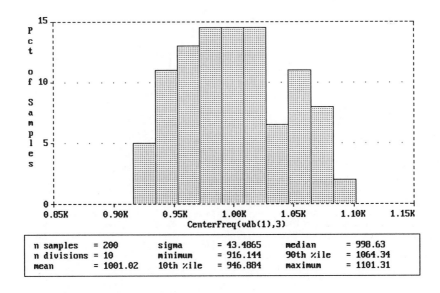

n samples	= 200	sigma	= 43.4865	median	= 998.63
n divisions	= 10	minimum	= 916.144	90th %ile	= 1064.34
mean	= 1001.02	10th %ile	= 946.884	maximum	= 1101.31

FIGURE 14-10 Output displayed as a histogram.

the individual deviations are specified, the overall deviation of the entire circuit **will become better approximated by the normal distribution**. This is not to say that you should not worry about each component specification "...since the overall result will be a normal distribution anyhow," as overall mean and variance may depend greatly on the individual distributions. These are the measures that will guide design decisions. However, you should not be surprised when most of your histograms look like bell curves. And being careful when specifying component variations will help make that bell curve an appropriate one.

Looking at the statistical metrics, in the box under the histogram, we notice that the mean and median are quite close to the design frequency of 1KHz. Good. And the standard deviation, or sigma (σ), is 43.5Hz or less than 5% of 1KHz. Also good! This means that, assuming this distribution of center frequencies is nearly normal, over 68% of these filters will operate within 5% of their designed center frequency using 10% components. We know this because the normal distribution curve has over 68% of its area within 1σ (one sigma) of its median. Also, the normal distribution curve has over 95% of its area within 2σ (two sigma) of its median, which means that 95% of the filters will operate within $2 \cdot 43.5\text{Hz} = 97\text{Hz}$ or 9.7% of the design frequency. We can readily verify this from the statistical metrics that show the tenth percentile at 946.9Hz, which is −5.3%, and the ninetieth percentile is at 1,064.3Hz, which is +6.4%. So we already know that 80% of our filters' center frequencies fall within the range of −5.3% to +6.4%.

Finally, we might want to get some measure of what percentage of the designs fall within certain ranges. To change the number of divisions in the histogram chart we

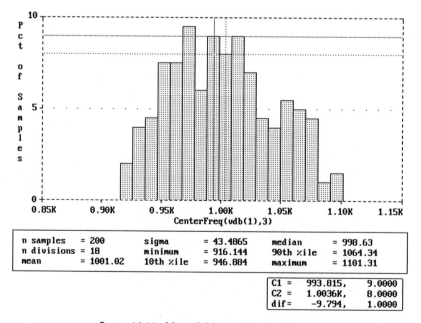

n samples	= 200	sigma	= 43.4865	median	= 998.63
n divisions	= 18	minimum	= 916.144	90th %ile	= 1064.34
mean	= 1001.02	10th %ile	= 946.884	maximum	= 1101.31

C1 =	993.815,	9.0000
C2 =	1.0036K,	8.0000
dif=	−9.794,	1.0000

FIGURE 14-11 More divisions and cursor display.

select the menu items X_axis, Options, and Histogram_divisions. By experiment, we find that 18 divisions yields a display with the divisions close to 50Hz in width and symmetrical about the design frequency of 1KHz. Then, by selecting the menu item Cursor, we use the cursors to measure the size of each division in the graph. All of this is shown in Figure 14-11, which captured the cursor readings for the two divisions closest to the design frequency of 9% and 8%. Adding these values would show, for example, that 17% of the distribution of center frequencies is within ±0.5% of the design frequency. This way you can analyze the distribution for whatever divisions interest you.

Now we will try the other filter circuit. In this case, we use the previous filter circuit but have replaced the inductor with a handful of components that implement a "gyrator."[†] This lets us make a resonant filter without using an inductor, which might be a constraint in an integrated circuit process or this circuit might be used where magnetic fields might couple into a normal inductor.

The circuit, shown in Figure 14-12, has the circuit file:

```
* RLC circuit using a gyrator
V1 in 0 ac 1
R1 in 1 RMOD 1K
R2 1  2 RMOD 100
R3 2  3 RMOD 100
R4 3  4 RMOD 100
R5 5  0 RMOD 100
.model RMOD res(R=1 dev/uniform 10%)
C1 1  0 CMOD 1.5915494u
C2 4  5 CMOD 1.5915494u
.model CMOD cap(C=1 dev/uniform 10%)
E1 2  0 (5,3) 1E6
E2 4  0 (1,3) 1E6
.ac LIN 801 600Hz 1.4KHz
.mc 200 AC V(1) MAX output(ALL)
.probe V(1)
.end
```

Note that the frequency range for the small-signal analysis is double that of the previous circuit because, as you will see, the variance in center frequency is greater than expected.

Running Probe, we select all the AC runs and the menu commands X_axis, Performance_analysis, and Add_trace to analyze the responses. Again we use the CenterFreq goal function to extract the center frequencies of each small-signal analysis. The resulting histogram is shown in Figure 14-13. Note that the number of divisions has been set so that each division accounts for about a 2% change from the design frequency.

A first glance at the histogram reveals that this circuit exhibits more variation than the previous circuit. The sigma (standard deviation) is over 60% larger and the span

---

[†] A gyrator is an impedance inverting device that converts an impedance Z into a reciprocal impedance $1/G^2Z$, where G is a constant. So a capacitor, with an impedance of $1/sC$, is converted to an impedance of sC/G^2, which corresponds to an inductance of C/G^2. For this circuit example $G = 1/R$, so $L = R^2C$.

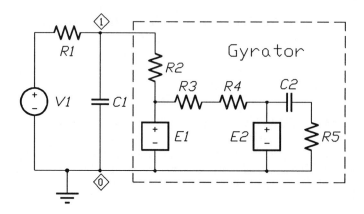

FIGURE 14-12   Filter circuit using a gyrator.

of minimum to maximum is almost twice the range of the simpler filter. With the tenth and ninetieth percentile metrics setting at nearly ±10% the design frequency of 1KHz, about 20% of the circuits are outside of this range. The previous circuit's variations were almost entirely within the ±10% range. There are other comparisons you might make to show that this circuit implementation might not be suitable.

So, take a gamble! By combining Monte Carlo methods with circuit simulation, you get another tool to simulate the variation of real circuits, provide insight into your circuit design, and for comparing different designs that provide the same function.

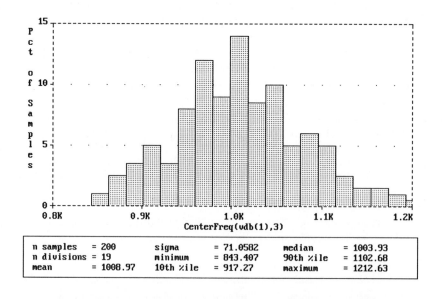

n samples	= 200	sigma	= 71.0582	median	= 1003.93
n divisions	= 19	minimum	= 843.407	90th %ile	= 1102.68
mean	= 1008.97	10th %ile	= 917.27	maximum	= 1212.63

FIGURE 14-13   Histogram from filter using a gyrator.

CHAPTER 15

SENSITIVITY ANALYSIS AND "WORST-CASE" PREDICTION

Sensitivity analysis and worst-case prediction are techniques that, like Monte Carlo simulation, are a smart use of "what if" approach to provide a practical result. Since the results of sensitivity analysis are used to "drive" the worst-case prediction, a sensitivity analysis is always done before a worst-case prediction. In fact, the two procedures are combined and always performed together.

15.1 SENSITIVITY ANALYSIS

Sensitivity analysis determines the rate of change in an output relative to a change in a circuit parameter. This is very similar to the DC sensitivity (.SENS statement) analysis described in Chapter 3, however the calculation is not limited to a DC result such as volts-per-ohm. By using collating functions, from Monte Carlo simulation, sensitivity information is generalized for other analyses, including AC and transient.

Since changing a component parameter may also change the DC bias-point of the circuit, PSpice may also recalculate the DC bias-point before working on the analysis you want. Components and parameters that affect bias, as well as another circuit specification (like gain), can be troublesome when it comes to trusting worst-case predictions. Often you will encounter a case where a component varies enough, or varies in concert with other components, to produce a circuit that behaves quite differently from the nominal design.

To predict worst-case operation, the nominal design is simulated with all the component values set to their nominal values. This is called the "nominal run" and it establishes the basis for subsequent sensitivity calculations and the initial value for testing the output of interest with the collating function. Then a simulation is performed for each component parameter that varies, and the small differences in the output of interest are saved. The final "worst-case run" is made only after all the individual variations have been simulated, and this final simulation is performed with each parameter skewed, in concert (for **better** or **worse** — your choice), which is why sensitivity analysis is part of worst-case prediction:

- Sensitivity is nothing more than a smart use of the "what if" idea. Individual sensitivities are calculated by a divided difference of a parameter value and an output measurement ("what if" we change this parameter just a little).

- Worst case is a smart use of the sensitivity information ("what if" we change **all** the parameters, together, to the **limit** of their tolerance) to provide the **worst** or **best** operation (your choice).

To predict extreme operation, PSpice must be told:

- how the circuit varies, and

- what "worst" or "best" means for the circuit.

15.2 SPECIFYING CIRCUIT SENSITIVITIES

To tell PSpice how the circuit can vary, all parameter variations are defined using device models. This means each device that you want to vary, even a resistor, will have a model.

As we saw with Monte Carlo simulation, some devices share model information, which means that you have several of the same types of device in your circuit. This is fine, and the variations in similar devices may also track each other (or "are correlated"). This can happen if the devices are manufactured at the same time, as happens with individual devices in an integrated circuit, transistor array, or a thin-film device. However, even the devices in an integrated circuit will show a small variation with respect to each other.

To handle these situations, two types of parameter variation are allowed: (individual) device variations and (as a group) lot variations. Either type of variation, or both, may be specified when describing how parameters vary. For example, you would use device variations to describe how discrete carbon-composition resistors are independently modeled, but you would use lot variations to describe the dominant variability in the elements of the thin-film R-2R ladder network in a digital-to-analog converter.

Unlike Monte Carlo simulation, sensitivity analysis is **not** concerned with the details of probability densities for the model parameters. What matters is which parameters vary and how they vary in concert with other parameters. Then, for "worst-case" prediction, **only** the extent of variation matters.

PSpice has two predefined probability distributions: the uniform distribution and the normal (or Gaussian) distribution:

- The UNIFORM distribution will vary over $\pm 1 \times$ the tolerance specified.

- The GAUSS (normal) distribution will vary over $\pm 3 \times$ the tolerance specified, since that value defines the standard deviation (or sigma "σ") for the distribution. Including the range of $\pm 3\sigma$ accounts for practical limit of 99.72% of the full distribution, which is theoretically infinite.

Any other distribution you may want needs to be defined, along with a unique name, using the .DISTRIBUTION statement. This statement creates a piecewise linear description of the probability distribution you want. The variation values are allowed over the range from -1 to $+1$, and the variation range is individually scaled for each parameter using it. (The .DISTRIBUTION statement is described in Appendix B.)

The piecewise description for parameter variations is useful for odd situations. For example, we may want to create a probability distribution for devices exhibiting an unsymmetrical variation of +10% and −20%. When creating the piecewise linear description we will want to map ±20% (the larger variation) onto −1 to +1 in the description: −20% will map to −1 and +20% will map to +1. Then we will set some of the probability distribution to zero to limit the positive variations to +10%. So, our new probability definition will look like:

```
.DISTRIBUTION plus10_minus20 (-1,0) (-1,1) (.5,1) (.5,0) (1,0)
```

The .MODEL statement that uses this probability distribution may look like:

```
.MODEL cheap_cap CAP(C=1 DEV/plus10_minus20 20%)
```

And the circuit device that uses this model may look like:

```
C17 gnd 17 cheap_cap .047u
```

The model parameter format is somewhat complicated by the range of options you might need to describe complex situations. Each model parameter in a .MODEL statement has the form:

name = value [[tolerance_spec]]

and it is the [[*tolerance_spec*]] that can be complicated. For this you will usually use a form like:

{DEV|LOT} *value*[[%]]

and use the default distribution, which is a uniform distribution (unless changed by the .OPTIONS statement). Including a percent sign "%" means the value is relative, as a percentage, and not an absolute value (in the same units as the parameter itself). The entire [[*tolerance_spec*]] has the following format:

[[DEV [[*track_&_dist*]] *value*[[%]]]]   [[LOT [[*track_&_dist*]] *value*[[%]]]]

which has identical sections for specifying device (DEV) and lot (LOT) deviation information: The device deviations are independent and the lot deviations track between devices that reference the same .MODEL statement. As mentioned before, you might think this stuff is only used for Monte Carlo simulation, however sensitivity analysis and "worst-case" predictions use the same information except the actual distribution of probability densities: Only the extent of each distribution is needed to create a "worst-case" prediction.

The optional [[*track_&_dist*]] information follows the keywords DEV or LOT with a slash "/" and **no spaces**, using the following format:

[[/*gen#*]]   [[/*distribution*]]

The *distribution* is simply a name, either predefined (UNIFORM and GAUSS) or one you have defined with a .DISTRIBUTION statement. The *gen#* is a digit, 0 through 9, that selects one of ten synchronized deviations between model parameters. You might think of this as defining a kind of lot tolerance between parameters in the same model, or different models, just like the regular lot tolerance between devices using the same model. By definition, there are only twenty synchronizers available to your circuit; ten for all DEV specifications and ten for all LOT specifications. Of course, if you do not use *gen#*, then the deviation is set independently.

As an example of why you might correlate deviations, let's say you were modeling an R-2R ladder with variability in the resistive material as well as variability in the contact resistance (where the resistors connect to conductors). The ladder is built using an integrated process, such as thick-film, thin-film, or an IC technique. Instead of using separate resistors in the circuit description to model the contact resistance, we can fold all the modeling into correlated models. The circuit description for the ladder elements might look like:

```
RA 25 26 R2mod 1  ; R2 leg
RB 26 27 R1mod 1  ; R1 leg
```

and so on. Note that the resistance specification has been set to 1, which will be multiplied by the model's R parameter to obtain the actual resistance. Now, the models will look like:

```
.MODEL R1mod res (R=10K DEV 1 LOT/3 5%)  ; note DEV = 1ohm
.MODEL R2mod res (R=20K DEV 1 LOT/3 5%)  ; note LOT = 5%
```

so that the models set the nominal resistance level, and an individual deviation of ±1 ohm for the contact resistance (assuming the default distribution type is UNIFORM). The lot deviation is ±5% for changes in the material that makes up the bulk of the device and the two models are synchronized in modeling their lot deviations, using the number "3" generator, which is as it should be for an integrated process. Synchronized deviations might prove useful for MOS circuit simulation, for example, where every MOS model's oxide thickness should be correlated.

15.3 CREATING A "WORST-CASE" PREDICTION

The .WCASE (or .WC) statement causes sensitivity analysis and worst-case prediction to be done, in due course, with other work by PSpice. There are four items PSpice needs:

- which analysis will be done, and you'll have a statement in the file describing that analysis,

- how to measure sensitivity and organize results,

- the kind and amount of output generated, and

- optional limits on how the sensitivities are managed.

.WCASE statement has the following form:

.WCASE {AC|DC|TRAN} *output_value function* [[*option...*]]

The *output_value* has the same specification as in the .PRINT or .PLOT statements. This output is "filtered" by *function*, which condenses the output from each analysis run, an entire waveform, into just one number! That number is used in the sensitivity calculations. The *function* specification is the same as for Monte Carlo simulation (see §14.6) where this is referred to as a "collating function." The available collating functions are:

- YMAX, which finds the greatest difference (plus or minus) along the output curve with respect to the nominal run's output.

- MAX, which finds the maximum (most positive) value along the output curve.

- MIN, which finds the minimum (most negative) value along the output curve.

- RISE_EDGE (*value*), which finds the first positive-going crossing through the threshold *value* along the output curve. Preceding the threshold there must be at least one analysis point below *value* and the reported output is the analysis point where the curve rises above *value*.

- FALL_EDGE (*value*), which finds the first negative-going crossing through the threshold *value* along the output curve. Preceding the threshold there must be at least one analysis point above *value* and the reported output is the analysis point where the curve falls below *value*.

The [[*option...*]] specification describes what kind of output will be generated, in the output file and for Probe, and limits the range over which the collating function operates. These options are described in Appendix B, but the most frequently used options are:

- LIST will print updated model parameter values used for the sensitivity analysis runs and the values used for the "worst-case" run.

- OUTPUT ALL requests output from the sensitivity runs, after the nominal (first) run. The content of the output from these runs is governed by the .PRINT, .PLOT, and .PROBE statements in the file. If OUTPUT ALL is omitted, then only the nominal and worst case (final) runs produce output.

- RANGE (*low*, *high*) restricts the range over which *function* will be evaluated. A "*" can be used to indicate "for all values." For example:

  · YMAX RANGE(*,.5) will evaluate YMAX for the waveform for values of the sweep variable (time, frequency, etc.) of .5 or less.

  · MAX RANGE(-1,*) will find the maximum of the waveform for values of the sweep variable (time, frequency, etc.) of −1 or more.

  If RANGE (*low*, *high*) is omitted, then *function* is evaluated over the whole sweep range. This is equivalent to RANGE (*, *).

- {HI|LOW} specifies which direction the worst case run is to go relative to the nominal run. If *function* is YMAX or MAX the default is HI; otherwise the default is LOW.

- VARY {DEV|LOT|BOTH} By default, any device that has a model parameter specifying either a DEV or a LOT tolerance will be included in the analysis. You may limit the analysis to only those devices that have DEV or LOT tolerances by specifying the appropriate option. The default is VARY BOTH.

The typical sensitivity analysis and worst-case prediction will be run with a statement like:

```
.WCASE AC VR(8) MAX RANGE(10K,*) HI LIST
```

and there will be a .PROBE statement in the file to collect output for Probe to display. This statement specifies that small-signal simulations will be analyzed, that the output VR(8) will be used for calculating sensitivity information (with MAX as the collating function, but evaluated only for frequencies from 10KHz up). The worst-case prediction will be run with variations set to produce the highest output (perhaps this is a "*best*-case" prediction) given the sensitivity calculations. For every run used for calculating sensitivities, the actual model parameter values used will be printed in the output file. Only the nominal run and "worst-case" run outputs will be available for any .PRINT, .PLOT and .PROBE commands in the circuit file. If the option OUTPUT(ALL) had been included in the .WCASE statement, then the runs for each sensitivity calculation would also produce output.

15.4 Special Care Required

"Worst-case" is a prediction and **not** an analysis. Nor is it an optimization process. Only the linear effect of each varying parameter is calculated, one parameter at a time, before the final simulation is run. It is **assumed** that the worst or best result will occur when all parameter values are shifted, simultaneously, to their respective limits in the direction indicated by the one-at-a-time sensitivities calculated for the nominal circuit. This prediction is guaranteed to be true **only** when the collating function provides a monotonic output over the range of all tolerance combinations. This is usually the case with normal, robust circuit designs. However, as we will see, you must be careful and check your results.

Trouble with "worst-case" occurs when your circuit is non-linear, which means just about any active circuit will have this problem — if you press hard enough! So let's try an example that is easy to visualize and for which we know the answer in advance to see how trouble can develop.

Our example is a non-linear amplifier, shown in Figure 15-1, which uses a controlled source that has a cubic polynomial transfer function of $-20x^3 + 10x$. (Dealing with non-linear amplifying devices may be familiar to older engineers that had to cope with tunnel diodes and unijunction transistors.) The circuit file for this

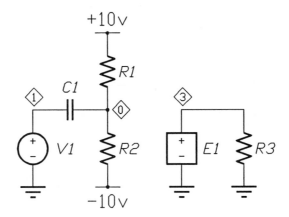

FIGURE 15-1 Simple amplifier circuit.

amplifier is:

```
* sensitivity and worst case
VP 10 0 DC +10
VN 11 0 DC -10

V1 1 0 AC 1
C1 1 2 1u
R1 10 2 RMOD 10.2K
R2 11 2 RMOD  9.8K
.model RMOD res(R=1 dev/uniform 5%)
E1 3 0 poly(1) 2 0 (0,+10,0,-20)
R3 3 0 50

.ac dec 10 10 10K
.wcase ac VR(3) MAX RANGE(10K,*) LIST HI
.probe
.end
```

The input to this amplifier is ac-coupled, through C1, and the amplifiying stage is biased to show what trouble can happen with "worst-case" prediction. The "worst-case" prediction is designed to find the combination of extreme values for the bias resistors that will produce a maximum output (or gain) at 10KHz. Note that the lower limit of the RANGE option is the upper limit of the small-signal analysis so the collating function measures at 10KHz only. Also, the real component of the output voltage is analyzed so we are not caught by phase reversals that would occur if the amplifying stage were biased at a cusp in its transfer function.

To plot the dc transfer function of the amplifying stage, in Probe, we can add the lines:

```
V2 2 0 DC 1
.dc lin V2 -1 1 .01
```

to the circuit file (this will spoil the other analyses so be sure to remove or comment out these lines, later). Now we can plot the transfer function of the amplifier and directly calculate the gain as the derivative of the output voltage with respect to the

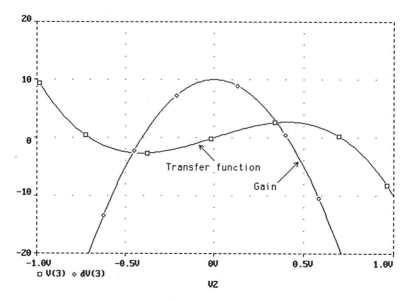

FIGURE 15-2 Transfer function and gain.

input voltage. These traces are shown in Figure 15-2.

By design, we have set the bias point to −0.2 volts, and simple analysis shows that if we can shift each of the bias resistors' values by the 5% tolerance limit, but in the opposing direction, the input bias point can be shifted about 0.5 volts. However,

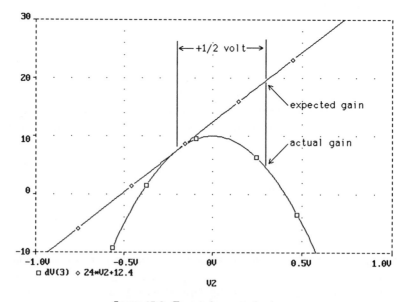

FIGURE 15-3 Expected vs. actual gain.

since the sensitivity calculations linearize the design problem, we can plot the effect of shifting the bias point, in advance. This is shown in Figure 15-3.

Running the "worst-case" prediction and plotting the results in Probe, as shown in Figure 15-4, we can verify that an inappropriate prediction has been made. True, the circuit gain did increase but a better value was attained when resistors varied less than their tolerance specification allowed.

Looking at the output file, we will see a section listing calculated sensitivities:

SENSITIVITY SUMMARY

```
*********************************************************************

    RUN             MAXIMUM VALUE

    R2 RMOD R       7.7183 at F =    10.0000E+03
                    (  15.579% change per 1% change in Model Parameter)

    NOMINAL         7.5999 at F =    10.0000E+03

    R1 RMOD R       7.4785 at F =    10.0000E+03
                    ( -15.973% change per 1% change in Model Parameter)
```

which lists the measured output change for a RELTOL change in model parameter value (0.1% in this example). The output change also shown normalized to a 1% change in model parameter value.

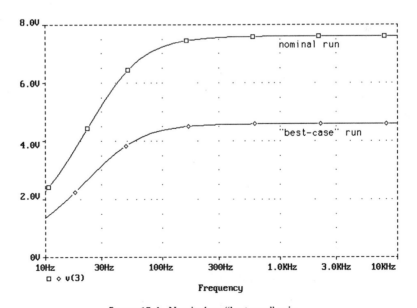

FIGURE 15-4 Nominal vs. "best-case" gain.

Further on in the output file, we will see a section showing the updated values the model parameters will take on to create the "worst-case" run:

```
****      UPDATED MODEL PARAMETERS      TEMPERATURE =   27.000 DEG C

**********************************************************************

        DEVICE          MODEL       PARAMETER      NEW VALUE
        R1              RMOD        R                  .95        (Decreased)
        R2              RMOD        R                 1.05        (Increased)
```

which shows how each model parameter will be changed for the final analysis run.

Finally, in the output file, we will see a section showing the measurements made for the nominal and worst-case runs:

```
                        WORST CASE SUMMARY

**********************************************************************

        RUN             MAXIMUM VALUE

    NOMINAL             7.5999 at F =    10.0000E+03

    ALL DEVICES         4.5891 at F =    10.0000E+03
                        (  60.384% of Nominal)
```

which shows what we knew already. This circuit's non-linearities will cause a "worst-case" prediction for **best** gain that is **less** than the nominal gain! So be careful.

Active devices, such as diodes and transistors, are at the heart of why SPICE was developed and became so successful. The behavior of these devices constrains the mathematics and numerics of the simulator, and only a few of the algorithms that could be used have proven successful in the face of the large changes in conductance that active devices undergo in normal circuit operation.

In this chapter we will take a qualitative tour of the models in PSpice, which are compatible with the devices in SPICE2 and SPICE3. The tables of parameters, with their names, descriptions, units, and default values, will **not** be in this chapter, as your SPICE manual will have these (and there are some in Appendix C of this book). Neither will you find the full set of equations for each device, which tend to obscure the external operation with which the user is familiar. Instead, we will focus on the terminal characteristics that electrical engineers know and show how the equations, with their copious parameters, describe and model these characteristics.

16.1 ACTIVE DEVICE MODELS

The semiconductor diode is usually the first active device one learns about; its ability to change resistance and switch, depending on current direction, is the basis for most beginning electronics courses. These courses, including semiconductor physics, will trace the development of the celebrated Shockley equation controlling *p-n* junction current, which is

$$\text{junction current} = I_{sat} \cdot \left(e^{\left(\frac{V_{junction}}{k \cdot T/q} \right)} - 1 \right) \tag{16-1}$$

It is sets of equations, like the Shockley equation, that defines the operation of active devices for SPICE. One of the options for the PSpice program is the source code for the routines containing these equations (albeit in a form that is efficient and suitable for the simulator) should you want to try different physics.

The parameters that are available through the .MODEL statement are ones that appear in the controlling equations for the device. For example, in the Shockley equation, the parameter IS (for I_{sat}) may be specified in the diode model. Of course, k and q are physical constants, and T is specified as the temperature for the simulation. In this way the user controls the device operation without writing new equations for each device.

Models for active devices are similar to those for passive devices, just more complicated in the conductances and currents that are calculated. For SPICE, it is not enough to consider, say, forward current gain for the bipolar transistor as an isolated feature of that device. All of the operating characteristics must be combined into a unified model, since SPICE is not capable of knowing when to discard effects that, for the circuit condition at hand, are negligible (this, of course, is a time-honored engineering practice). All of the characteristics that affect the calculation of conductance, transconductance, current, et cetera, must be present each time the device is evaluated. This means that device operation, which we normally split into operating regions, such as *saturation* and *cutoff*, becomes one continuous set of formulas. It is difficult to develop device models that behave in this way.

The benefit, for the SPICE user, is that all of the device characteristics can be included in the simulation. Of course, you may choose to ignore some characteristics. Often you find that a circuit does not simulate the way you expect because it has "been had" by some device characteristic that you overlooked during design. This is the purpose of SPICE: to verify the operation of a circuit (after all, the simulator is dumb and will not be misled by what you intended the circuit to do). This is why you want models that are complete enough to not only simulate your circuits when they behave as you expect, but also to show you when they don't. So, the models are important.

There is only one model, in SPICE, for each device type. This model is the full nonlinear set of equations describing currents, conductances, and capacitances. New SPICE users often wonder if, for small-signal analysis, the simulator uses a "hybrid-π" model for the bipolar transistor. It does, but it is imbedded in the nonlinear equations, which are used to arrive at the bias-point for the circuit and then the conductance, transconductance, and capacitance values are saved for use by the small-signal analysis section of the simulator. Think of it as PSpice calculating a hybrid-π model for each transistor in the circuit. But the topology internal to the transistor is the same for transient and small-signal analysis. For the latter analysis, the small-signal values are calculated and used.

16.2 SEMICONDUCTOR DIODE

The diode model in PSpice, as mentioned before, contains a nonlinear current source that follows the Shockley equation:

$$\text{current} = \text{IS} \cdot \left(e^{\frac{Vj}{N \cdot Vt}} - 1 \right) \tag{16-2}$$

where

Vj is the voltage across the junction

Vt is the thermal voltage $(= k \cdot T/q)$

These values, with the model parameters IS and N, are used to model the current-voltage effects of the semiconductor junction. This does not include the nonideal operation of real diodes. For example, at low currents (less than 1nA), other semiconductor processes that increase the flow of currents become noticeable. As a practical matter, these are small currents that are ignored by SPICE.

As you can see in Figure 16-1, by setting IS to different values you can obtain the characteristics of (i) a Schottky-barrier diode, and (ii) a silicon diffused-junction diode. High-current effects are modeled, grossly, by including a series resistance that is intended to combine the effects of bulk resistance (the material on each side of the junction) and high-level injection. At high currents, the observed diode current stops following the Shockley form

$$I_{forward} = \text{IS} \cdot e^{\frac{Vj}{N \cdot Vt}} \qquad (16\text{-}3)$$

and approaches a modified form

$$I_{forward} = \text{IS} \cdot e^{\frac{Vj}{2 \cdot N \cdot Vt}} \qquad (16\text{-}4)$$

Again, for practical reasons (and because of an emphasis on integrated circuits, which rarely develop such large currents in normal operation), SPICE does not include this modified form. Instead, SPICE has only the series resistance parameter, RS, available for more limited modeling of this effect. It is not so important to model the effect accurately as to make a provision for the effect so it is available to indicate abnormal operation. However, PSpice **does** include this effect and uses the parameter IK to determine the onset of high-level injection.

As you can see in Figure 16-2, by setting RS you can limit the exponential effect of the Shockley equation and the device becomes resistive.

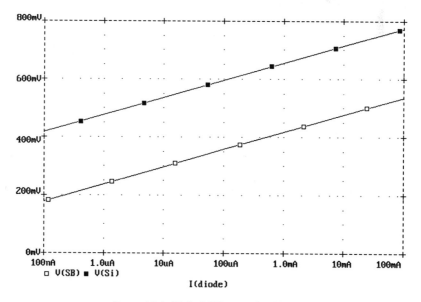

FIGURE 16-1 Typical I-V curves for diodes.

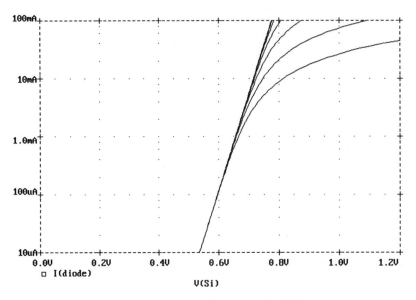

FIGURE 16-2   Diode I-V curves, varying RS.

For reverse operation, the value of IS, which the Shockley equation asymptotically approaches, is usually too small a value as real devices have leakages that allow current across the junction. To help model this, as well as improve the operation of the simulator, a minimal conductance is connected in parallel with the junction. The value of the conductance is set by the option GMIN (see the .OPTION statement in Appendix B, page 231).

As you can see in Figure 16-3, the reverse diode current deviates from the Shockley equation due to the GMIN conductance in parallel with the junction. Reverse breakdown, as found in Zener diodes, is modeled by another exponential form:

$$I_{reverse} = \text{IBV} \cdot e^{\dfrac{-(Vj+\text{BV})}{Vt}} \qquad (16\text{-}5)$$

which gives the correct, but not exact, effect of reverse breakdown. One problem is that semiconductor junctions have more than one breakdown mechanism, and these processes can occur at the same time. Again, as a practical matter, SPICE does not attempt to model the "blend" of processes; the simple form serves most engineering purposes. However, PSpice **does** provide more accurate modeling of reverse breakdown, using several parameters to implement an empirical formula that closely matches the Zener voltage and impedance of standard devices.

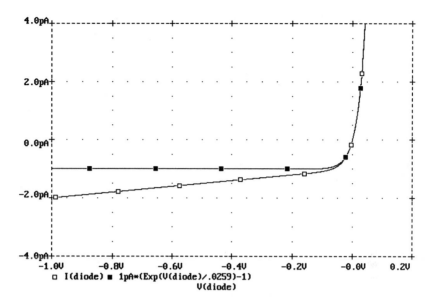

FIGURE 16-3 Diode reverse current and Shockley equation.

Figure 16-4 shows both forward and reverse operation. Note that this figure uses extraordinary parameter values to exaggerate the differences from the Shockley equation.

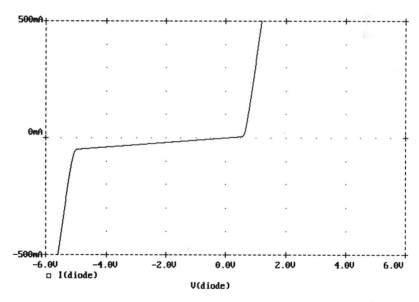

FIGURE 16-4 "Full range" plot (exagerated values).

Diode capacitance is modeled by a voltage-dependent capacitor, which is connected in parallel with the nonlinear current generator described previously, to represent the charge storage effects of the junction. There are two components to this charge:

- the reverse-voltage capacitive effect of the depletion region, and

- the forward-voltage charge represented by mobile carriers in the diode junction.

Reverse-voltage capacitance follows the simple approximation that the depletion region (the area of the junction that is depleted of carriers) serves as the gap between the "plates" of a capacitor. This region varies in thickness, and therefore the capacitance varies with applied voltage. For a step (abrupt) junction, or linearly-graded junction, the capacitance approximation is

$$\text{capacitance} = \frac{\text{CJO}}{\left(1 - \dfrac{V_j}{\phi}\right)^{\text{M}}} \qquad (16\text{-}6)$$

where CJO is the zero-bias value, ϕ (phi) is the junction barrier potential, and M is the grading coefficient that varies (1/2 is used for step junctions and 1/3 is used for linearly-graded junctions, and most junctions are somewhere in between).

There is often confusion about the barrier potential, which appears in the capacitance equation. From capacitance measurements, ϕ (model parameter VJ, and not to be confused with the V_j in the equations) takes on a value of nearly 0.7 volt for regular (silicon) junction diodes and a range of 0.58 to 0.85 volt for various Schottky-barrier diodes. This value sometimes is confused with either the forward-current voltage drop of the diode or the energy gap of the material; it is **neither** of these.

As you can see in Figure 16-5, varying M will generate a variety of reverse-bias capacitance characteristics. Some inspection of the capacitance formula reveals that it predicts infinite capacitance for a forward bias, which is not the case for a real junction. Several depletion capacitance formulas have been proposed that more correctly fit observed operation; however, SPICE uses a simple approach: for forward biases beyond some fraction (set by the parameter FC) of the value for ϕ, the capacitance is calculated as the linear extrapolation of the capacitance at the departure. This provides a continuous numerical result, and does not affect circuit operation significantly because, for forward bias, the device capacitance is normally dominated by diffusion capacitance.

The diffusion charge (and therefore the capacitance) varies with forward current and is simply modeled as a *transit time* (model parameter TT) for the carriers to cross the diffusion region of the junction. The total charge is

$$\text{diffusion charge} = \text{device current} \times \text{transit time} \qquad (16\text{-}7)$$

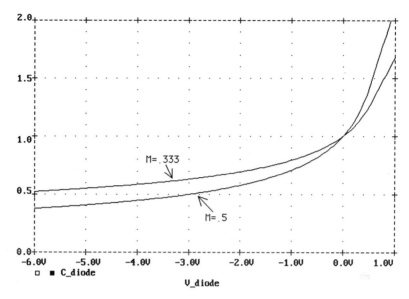

FIGURE 16-5 Junction capacitance (relative).

and capacitance is the derivative, with respect to bias, of this:

$$\text{diffusion capacitance} = TT \cdot \frac{IS}{N \cdot Vt} \cdot e^{\frac{Vj}{N \cdot Vt}} \qquad (16\text{-}8)$$

Diffusion charge manifests itself as the *storage time* of a switching diode, which is the time required to discharge the diffusion charge in the junction, which must happen before the junction can be reverse-biased (switched off). Storage time is normally specified as the time to discharge the junction so that it is supporting only a fraction (typically 10%) of the initial reverse current. First, a forward current is supplied to the device to charge the junction. Then, as quickly as possible, a reverse current is supplied to the device. Internally, the junction is still forward-biased to a voltage nearly the same as before the switch in current; the junction is still conducting at the forward current rate. This internal current adds to the external current as the total current discharging the junction. As the junction voltage decreases, the internal current falls off exponentially (according to the Shockley equation). This system is a relatively simple differential equation that can be solved to an explicit equation for the TT parameter (assuming complete discharge) as follows:

$$\text{transit time} = \frac{\text{storage time}}{\ln\left(\dfrac{I_F - I_R}{-I_R}\right)} \qquad (16\text{-}9)$$

As you can see from Figure 16-6, the diffusion charge dominates the reverse recovery characteristic of the diode. During the last part of the recovery, as the junction becomes reverse-biased, the depletion capacitance dominates. This causes the small

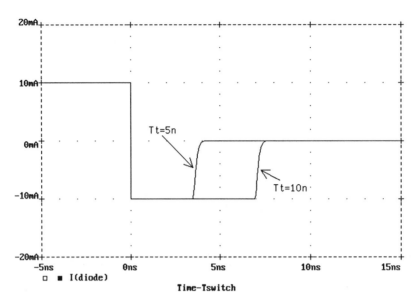

Figure 16-6 Reverse recovery in time domain.

tail at the end of the discharge cycle. Total capacitance is taken to be the sum of these capacitances: the depletion approximation dominates for reverse bias as the device current is small, and the diffusion approximation dominates for forward bias as the device current is large.

16.3 Junction Field-Effect Transistor (JFET)

The JFET is the simplest of the transistor devices. In this device, the increase in the depletion region by gate junction bias "pinches" the channel, increasing its resistance to drain current. It is known as a *square law* device because of the expression relating drain current to gate-to-source voltage:

$$I_{drain} = \beta \cdot \left(V_{GS} - V_{threshold} \right)^2 \tag{16-10}$$

This equation is an approximation of the transfer function given by the exact analysis of channel charge, but is almost universally used (see Figure 16-7). Another way of arriving at the same square law relation is by making the approximation that the gate junction capacitance is a linear function of the gate junction voltage (which in turn describes how the channel region is modulated). As we saw for the diode, reverse-bias capacitance is not a linear function, but it may be approximated that way for biases much larger than the barrier potential (ϕ) of the junction. The error associated with using the square law form happens to be quite small (when compared to exact analysis as well as real devices). The square law result applies only when V_{DS} is greater than $V_{GS} - V_{threshold}$ (where $V_{threshold}$ is equivalent to the *pinch-off* voltage of the JFET, and set by parameter VTO), when the channel of the FET is *saturated*.

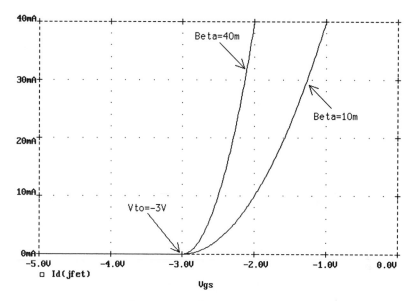

FIGURE 16-7 JFET drain current.

When V_{DS} is below pinch-off, the expression relating drain current to gate junction voltage is

$$I_{drain} = \beta \cdot \left(2 \cdot \left(V_{GS} - V_{threshold}\right) \cdot V_{DS} - V_{DS}^{2}\right) \qquad (16\text{-}11)$$

which describes (on an I_{drain} versus V_{DS} plot) an inverted parabolic curve passing through the origin and which, at its peak value (when V_{DS} is at pinch-off), intersects the square law formula. This parabolic region of operation is called the *linear* region; for small drain voltages, the expansion of the equation (above) is dominated by the linear term

$$I_{drain} \approx \beta \cdot 2 \cdot \left(V_{GS} - V_{threshold}\right) \cdot V_{DS} \qquad (16\text{-}12)$$

Finally, I_{drain} is zero when V_{GS} is less than $V_{threshold}$ (see Figure 16-8).

Real JFETs, in the saturation region, are not ideal current devices, since their drain currents vary with drain voltage. This effect is modeled by the device parameter LAMBDA, which sets the output conductance

$$I_{drain} = \text{BETA} \cdot \left(V_{GS} - \text{VTO}\right)^{2} \cdot \left(1 + \text{LAMBDA} \cdot V_{DS}\right) \qquad (16\text{-}13)$$

which yields an increasing drain current for increasing values of V_{DS}, as shown in Figure 16-9.

Since transconductance is

$$\frac{\partial I_{drain}}{\partial V_{GS}} \qquad (16\text{-}14)$$

we can readily relate BETA to transconductance by differentiating the drain current

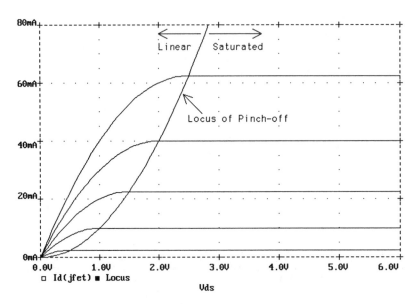

Figure 16-8   JFET drain current "curve family."

formula to get

$$\text{transconductance} = 2 \cdot \text{BETA} \cdot \left(1 + \text{LAMBDA} \cdot V_{DS}\right) \cdot \left(V_{GS} - \text{VTO}\right) \qquad (16\text{-}15)$$

The capacitances of the JFET follow the form we saw for the diode. Both the gate-to-source and gate-to-drain junctions have a nonlinear capacitor. The zero-bias capacitance value is selected for each junction. When these junctions become forward-biased, the straight-line extension of capacitance is used (just like the diode). However, there is no provision for diffusion charge in the junction, since JFETs are rarely used in a mode that has either junction forward-biased.

16.4 Gallium-Arsenide MESFET (GaAsFET)

The GaAsFET is a Schottky-barrier gate FET, or MESFET (for METAL and SEMICONDUCTOR FET), made of gallium arsenide ("GaAs" is its chemical abbreviation). The primary advantage of GaAs over silicon is its electron mobility, that is six times greater (mobility is the speed of electrons in the material for a given electric field that propels the electrons). This is an advantage of GaAs that is important for high-frequency electronics. At present, the major drawback to GaAs is the difficulty in processing and manufacturing devices; however, these problems are being solved quickly due to a large market for high-speed devices in computing and defense electronics. Soon the fastest computers, using GaAs devices in modules designed using PSpice simulations, will be simulating tomorrow's circuits.

The GaAs MESFET operation is like the silicon MOSFET. An insulating layer between the gate and channel is provided by the potential barrier formed at the

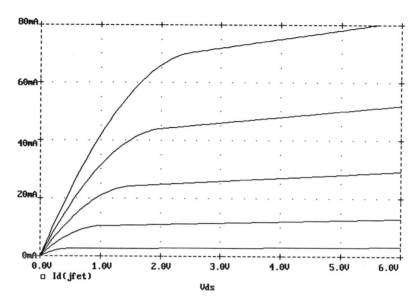

FIGURE 16-9   JFET curves showing output conductance.

contact of two materials, in this case a metal gate and GaAs substrate. Similar to the MOSFET a channel charge is induced to create a conducting path under the gate, connecting the drain and source of the device. However, the detailed device operation is different in that, in GaAs, the electron velocity *saturates* for an electric field roughly ten times lower than in silicon. Thus the saturation in drain current, for GaAs, occurs due to carrier-velocity saturation, whereas channel pinch-off causes this in silicon. There are several proposed models for the conductivity of the channel, so PSpice includes both the *Curtice* model and the *Raytheon* model.

The Curtice model was one of the first to be implemented in a circuit simulator, circa 1983, and therefore gained early acceptance. The formula for active operation, where V_{GS} is greater than $V_{threshold}$, is

$$I_{drain} = \beta \cdot \left(V_{GS} - V_{threshold}\right)^2 \cdot \left(1 + \lambda \cdot V_{DS}\right) \cdot tanh\left(\alpha \cdot V_{DS}\right) \qquad (16\text{-}16)$$

which includes both *linear* and *saturated* modes and is an empirical fit using the hyperbolic tangent function (see Figure 16-10). The device capacitances for the Curtice model are simple: the normal *p-n* junction capacitance is used for the gate-source and gate-drain capacitance, and a fixed capacitance is available for drain-source capacitance modeling.

The Raytheon model (named after the employer of the developers) is a more recent model, circa 1986, and benefits from the later research into GaAs devices. In particular, it has two improvements over the Curtice model:

• an enhanced drain current formulation, and

• a new capacitance model.

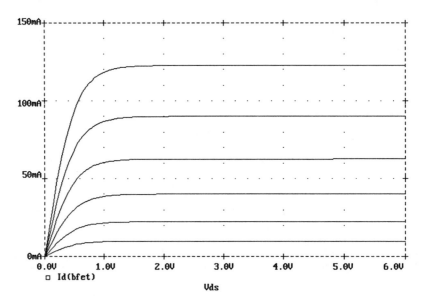

FIGURE 16-10 GaAsFET "curve family" for Curtice model.

The drain current formula was modified; while the Curtice model did well for the observed change in drain current versus V_{DS}, it did not do well for drain current versus V_{GS}. Using observed operation, the portion of the Curtice formula

$$I_{drain} = \beta \cdot \left(V_{GS} - V_{threshold}\right)^2 \tag{16-17}$$

was changed to

$$I_{drain} = \frac{\beta \cdot \left(V_{GS} - V_{threshold}\right)^2}{1 + B \cdot \left(V_{GS} - V_{threshold}\right)} \tag{16-18}$$

to effect the desired operation for $V_{GS} \gg V_{threshold}$. To increase computational efficiency, the hyperbolic tangent function was replaced with a polynomial approximation

$$tanh(x) \approx \begin{cases} 1 - (1 - x/3)^3 & \text{for } 0 < x < 3 \\ 1 & \text{for } x \geq 3 \end{cases} \tag{16-19}$$

Note that $x = 3$ defines the onset of device saturation. Using Figure 16-11, compare the Raytheon model to the Curtice model.

The new capacitance model comes (again) from the problem of carrier-velocity saturation. In theory, the gate-source capacitance should increase abruptly at the onset of velocity saturation and the gate-drain capacitance should decrease abruptly. In practice, velocity saturation occurs more gradually so the capacitance changes will not be as abrupt. The Raytheon capacitance model is "charge oriented" to calculate the effects of velocity saturation. Once the channel charge, and therefore capacitance

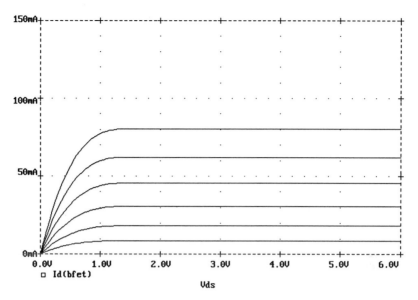

FIGURE 16-11 GaAsFET "curve family" for Raytheon model.

(as the change in charge versus voltage), is calculated, it is split into gate-source and gate-drain capacitance values. Furthermore, these capacitance values maintain symmetry if the device is operated in the inverted mode.

16.5 BIPOLAR JUNCTION TRANSISTOR (BJT)

The bipolar junction transistor, or BJT, model in PSpice is an enhanced version of the Gummel-Poon model. This means that it is also a superset of the earlier Ebers-Moll model, as well as its more basic form, which is usually the first model encountered by an electrical student. You have access to all of these levels of model by the way the Gummel-Poon parameters are set, or defaulted. Associated with this DC model are all of the junction capacitances, which, with some care, give good small-signal and transient simulation results up to microwave frequencies.

Both the Ebers-Moll and Gummel-Poon models are symmetrical, with both forward and reverse operation (just like a "real" bipolar transistor). Therefore, there are forward and reverse parameters that are explicitly labeled as such; however, some of the parameters labeled as being associated with the base-emitter or base-collector junctions are also forward or reverse parameters (respectively). This means that of the forty-odd parameters in the bipolar model, most of them are duplicates specifying reverse operation, or base-collector instead of base-emitter characteristics. So, the list of parameters is not as formidable as it looks.

Using graduated models is a common way to teach transistor theory; this is easy to do, since these models trace the development of the theory of transistor operation. First came the simple, nonlinear model described by Ebers and Moll, in 1954, which

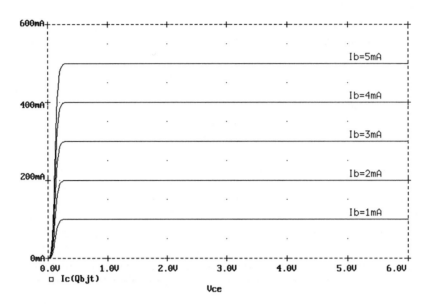

Figure 16-12 Collector current "curve family."

was a DC model only (that is, it did not include capacitive effects). This is the model you get in SPICE using the default values for the BJT parameters; the forward beta (BF) is 100, the reverse beta (BR) is 1, and IS is set to provide a normal base-emitter voltage for a small device (default model curves are shown in Figure 16-12).

To get to the next level of model, you would include the junction capacitances and parasitic resistances for each of the terminals. The capacitance models are identical to the ones we saw earlier for the diode, applied to both the base-emitter and base-collector junctions; these provide correct transient/frequency response and include diffusion charge to model switching times correctly. This includes the depletion capacitance parameters:

- CJE and CJC, which are equivalent to CJO for the diode;
- MJE and MJC, which are equivalent to M for the diode;
- VJE and VJC, which are equivalent to VJ for the diode;
- FC (for b-e **and** b-c junctions), which is equivalent to FC for the diode.

This also includes the diffusion capacitance parameters, TF and TR, which are equivalent to TT for the diode.

The parasitic resistances model the bulk resistance included in the physical construction of the device. The emitter (RE) and collector (RC) resistances alter the terminal characteristics, decreasing the slope of collector current for low collector-emitter voltages. The base (RB) resistance primarily affects frequency response and noise.

The final level of model includes carrier recombination and base-width modulation effects that provide the realities of gain variation. These effects are associated with the Gummel-Poon model, although included in enhanced versions of Ebers-Moll models, because the Gummel-Poon model treats a number of effects in a unified manner. The terminal characteristics are not that much different between the Gummel-Poon and earlier "enhanced" Ebers-Moll models, but the underlying physics is better. The SPICE user must remember that most of simulation modeling is a curve fitting game; a variety of approaches will give the same (for engineering purposes) simulation results.

Base-width modulation comes from the voltage across the base-emitter and base-collector junctions. The most obvious effect is a finite output conductance, or an increase in collector current with base-collector voltage, which was called the "Early" effect (after J. M. Early, who first reported the phenomenon). The forward parameter is the Early voltage, VAF (or VA), from the following geometric interpretation: extrapolating the collector currents, in saturation, forms a converging set of lines that intersect the negative X-axis at the Early voltage (which, however, is expressed as a positive value). The output conductance, $1/h_{oe}$, is the slope of the extrapolated lines.

There is also a reverse-Early voltage, VAR (or VB), sometimes called the "Late" voltage. This parameter has the same effect, and geometric interpretation, for reverse transistor operation (see Figure 16-13).

Carrier recombination and leakage accounts for the decrease in current gain at low current levels. Not all of the current flowing through the base terminal is available for transistor action; some of it leaks off and some is lost to recombination. Only the current that escapes these effects participates in the amplification action of

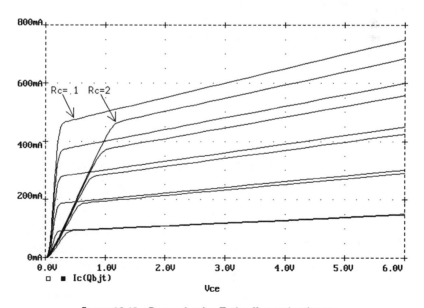

FIGURE 16-13 Curves showing Early effect and resistance.

the transistor. Leakage and recombination currents have voltage dependencies similar to the Shockley equation:

$$\text{leakage current} = \text{IS}_\text{L} \cdot \left(e^{\frac{Vj}{4 \cdot Vt}} - 1 \right)$$

(16-20)

$$\text{recombination current} = \text{IS}_\text{R} \cdot \left(e^{\frac{Vj}{2 \cdot Vt}} - 1 \right)$$

These currents are part of the base junction current, and similar currents occur in other semiconductor junctions. For example, the semiconductor diode also has these currents, but they are not modeled in SPICE because the effects do not matter for most circuits; they do matter for more complete modeling of the bipolar transistor.

As a simplification, the leakage and recombination currents are combined into a single formula:

$$\text{composite ("lost") current} = \text{ISE} \cdot \left(e^{\frac{V_{BE}}{\text{NE} \cdot Vt}} - 1 \right)$$

(16-21)

where ISE and NE are the values used that make the composite formula match the combination of the previous formulas. The composite formula is the Shockley equation for a nonideal, or *leakage*, diode.

In the Gummel-Poon model, this leakage diode is connected in parallel with an ideal diode to represent the base-emitter junction. The current through the ideal diodes takes part in the transistor action (its current is multiplied by beta to generate collector current); the leakage diode current does not. The parameters IS/BF and NF are the saturation current and emission coefficient, respectively, for the ideal diode. The parameters ISE and NE are the saturation current and emission coefficient, respectively, for the leakage diode. The formula for junction current, for each diode, is

$$\text{ideal diode current} = \frac{\text{IS}}{\text{BF}} \cdot \left(e^{\frac{V_{BE}}{\text{NF} \cdot Vt}} - 1 \right)$$

(16-22)

$$\text{leakage diode current} = \text{ISE} \cdot \left(e^{\frac{V_{BE}}{\text{NE} \cdot Vt}} - 1 \right)$$

As mentioned earlier, these currents also occur in the semiconductor diode. If you wanted to model a diode more accurately in the low-current, forward region, then you might try using this dual-diode approach.

Forward current gain is defined as the ratio of collector current to base current. With the leakages in the nonideal diode having an emission coefficient, NE, around 2 (the ideal diode's emission coefficient, NF, is usually unity) the percentage of total current "leaking" increases with decreasing current (both diodes have the same voltage across their terminals). The remaining current available for transistor action decreases with decreasing current and along with it the apparent beta. Here, beta (the BF parameter) is constant but the amount of current in the ideal diode is decreasing.

Forward current gain will be reduced to half of BF when these two currents are equal, and, since they are in parallel the junction voltages are identical:

$$\text{``half--beta'' current} = \left(\frac{ISE^{NE}}{(IS/BF)^{NF}} \right)^{\frac{1}{NE-NF}} \tag{16-23}$$

and the asymptotic slope of the reduction in forward gain, with decreasing current, is

$$\frac{\partial \text{beta}}{\partial ln(I_C)} = 1 - \frac{NF}{NE} \tag{16-24}$$

Curve fitting is used to match device measurements to these equations.

We can explore the effects of ISE and NE by using Probe to display base and collector currents. If we sweep an injected base current (using the DC sweep) and display the logarithm of base and collector current with respect to the logarithm of the ideal diode current, we can see how forward beta changes over a broad range of forward operation.

Sweeping the base current and the value for ISE, we get Figure 16-14. In this figure the X-axis has been set to the value of the collector current. Now we can plot I_C and I_B, then label the intercepts of the trace to indicate various model parameters. Note that both axes are logarithmic. The vertical distance between the base and collector currents is the logarithmic value of the DC beta. So we can see that ISE sets the onset of reducing beta.

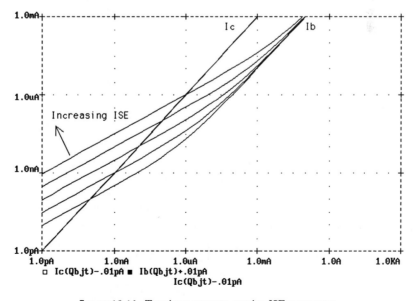

FIGURE 16-14 Transistor currents, varying ISE parameter.

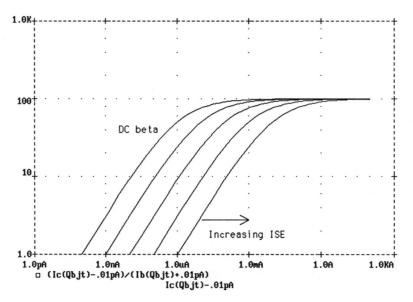

FIGURE 16-15 Transistor DC beta, varying ISE parameter.

To see this more clearly, in Figure 16-15 we display DC beta directly as I_C/I_B. The onset of beta reduction is very clear, but it is now more difficult to relate the curves to the model parameters IS and ISE. Similarly, we may sweep the base current and the value for NE (the emission coefficient). Again, the X-axis has been set to I_C but with a correction factor; specifically, the diodes that model the reverse characteristics of the transistor have currents that must be accounted for. Similarly, the traces for I_C and I_B have corrections. This is done in Figure 16-16 as the traces are shown down to extremely low currents where these corrections are significant. The traces for base current converge at the value of the model parameter ISE with an asymptotic slope of 1/NE. Again, the distance between the base and collector currents is the logarithm of the DC beta. The value of NE sets the rate at which DC beta decreases with decreasing collector current.

Of course, similar parameters are available for reverse operation: IS/BR is the reverse saturation current for the reverse "ideal" diode; ISC and NR are the saturation current and emission coefficient, respectively, for the reverse leakage diode.

Finally, the base-width modulation effect also accounts for the reduction in current gain with increasing collector current, a mode called *high-level injection*. Charge conservation reduces the efficiency of the transistor action, with high-current beta having a dependence on collector current of

$$\text{forward beta} = \frac{\text{BF}}{1 + I_C/\text{IKF}} \qquad (16\text{-}25)$$

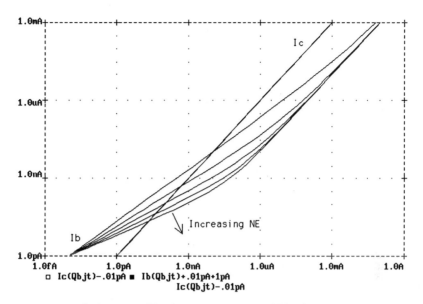

FIGURE 16-16 Transistor currents, varying NE parameter.

where IKF is the forward "knee" current. Solving this formula to find the collector current yielding beta equal to BF/2 shows that this occurs at a collector current equal to the value of IKF. This time the X-axis has been set to V_{BE}; this is proportional to the logarithm of the base (ideal diode) current and is therefore proportional to the logarithm of the "ideal" collector current (which we used in the previous figures). If we had set the X-axis to I_C, the trace of I_C would be a straight line, as in the previous figures and we would not see the deviation of I_C due to high injection.

Since DC beta is the ratio of I_C to I_B, the vertical distance between I_C and I_B in Figure 16-17 is the logarithm of DC beta. We can see that at low currents, DC beta is unity (where the traces cross) and increases to a maximum, then decreases to unity at high currents (where the traces cross again). IKR is available to model reverse operation in the same way.

This completes our internal model of the bipolar transistor for Gummel-Poon, although further improvements have been made to extend the usefulness of the model. In particular, the model includes:

- variable base resistance, to provide for base current "crowding" effects,

- split base-collector capacitance, to model more accurately high-frequency response, and

- a variable, forward transit time to reduce frequency response at high collector currents.

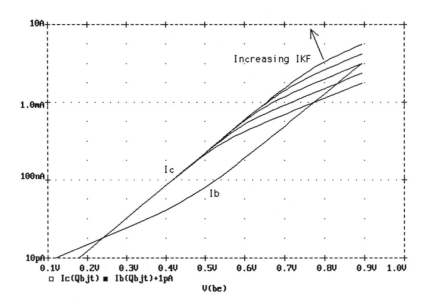

Figure 16-17   Transistor currents, varying IKF parameter.

16.6 MOS Field-Effect Transistor (MOSFET)

The MOS modeling techniques of SPICE2 (and PSpice) were a significant improvement over those in the original SPICE, as well as most other simulators. The MOS models, and the availability of SPICE2 in the public domain, made the simulator popular with integrated circuit designers worldwide; these two features probably clinched the *"de facto* standard" title for SPICE2. But being widely used does not mean the models were widely appreciated. A cottage industry grew up surrounding SPICE2 primarily to modify the MOS models and their interaction with the circuit-solution algorithms. Nearly every integrated circuit manufacturer has an employee, or group, which supports an internal version of SPICE2. Whereas SPICE2 added two levels of MOS model to the original SPICE, some commercial programs, such as HSPICE, have an additional eighteen levels of MOS model.

This dissatisfaction with the U. C. Berkeley MOS models stems, in part, from a lack of documentation of the models themselves. There is one laboratory report, from U.C. Berkeley, which describes the models (as planned) and, of course, there is the model code (as built) — over 2,000 lines of FORTRAN that do the calculations. In many cases, these sources have not been enough to use the models successfully. We will cover the formulations and parameters required to specify a model, without going into the mathematics that are covered in the U.C. Berkeley report (see the references at the end of this chapter).

Currently, the SPICE3 simulator provides five built-in "levels" of model, which are also available in PSpice:

Level 1: "Schichman-Hodges" is a basic MOSFET model suitable for preliminary analysis. Its equations are similar to the JFET.

Level 2: "analytic model" is a geometry-based model developed from one-dimensional device physics and includes moderate two-dimensional corrections. This model has limitations for smaller device geometries.

Level 3: "semi-empirical model" is a more qualitative model that uses (as the name would imply) observed operation to define its equations.

Level 4: "BSIM" advances the semi-empirical approach, with over sixty parameters to describe electrical operation and small-geometry effects.

Level 5: "BSIM3" advances the BSIM approach, addressing the challenges of VLSI circuit design in advanced submicron and sub-half-micron technologies.

The features of these models are summarized in Table 16-1.

The traditional arguments over MOS models usually happen over levels 2 and 3; level 1 is normally used for large devices (discrete parts, such as signal MOSFETs and power MOSFETs) or for the "first pass" at an integrated circuit design to check that the circuit is connected and functionally correct.

The MOS models were set up for considerable flexibility in the use of parameters. While conceptually it is convenient to separate each model from the others, the equations in the code do not make this distinction for the entire model. For each characteristic, a selection of ways is available for calculating the model: level 1 is elementary, level 2 uses processing parameters and geometry, level 3 uses measured characteristics. For example, the level 2 method calculates threshold voltage from the specification of doping concentration, surface state density, et cetera, so the overall model accuracy depends on having good formulas and accurate data. However, the level 3 method uses a measured value for the threshold voltage, so the overall model accuracy depends on the ability of the engineer to match the characteristics of the

TABLE 16-1 Comparison of MOS model features.

Features	Level 1 _developed 1968_	Level 2 _developed 1980_	Level 3 _developed 1980_	Level 4 _developed 1985_	Level 5 _developed 1992_
Technology	$\geq 4\mu m$	$\geq 1.2\mu m$	$\geq 1.0\mu m$	$\geq 0.8\mu m$	$\geq 0.3\mu m$
I_D Parameters	8	23	21	67	21
Small-Geometry Effects	NO	Complex Expressions	Semi-empirical expressions	Globally fixed format	Locally adapted format
Subthreshold Effects	NO	YES	YES	YES	YES
Capacitance Effects	Meyer model	Conservation	Meyer model	Conservation	Conservation
Temperature Effects	Limited	YES	YES	External	YES
Noise Effects	NO	YES	YES	NO	YES

component with the parameters of the model. You can use a mix of methods; for example, you could choose the level 2 technique for calculating threshold voltage, and the level 3 technique for calculating drain current. It would be more accurate to say that there are three methods, instead of models, as all of the combinations of the modeling sections provide many unique paths for calculating device characteristics.

Which method is used is determined, in part, by the model parameters specified. For example, if the substrate doping is specified, the analytic model will be used for some calculations regardless of level. This selective calculation makes it possible to use the empirical model even though the parameter based on measured data is not available, by calculating the parameter from process data. This approach tries to arrive at a consistent set of parameters for the model equations. When enough parameters are not supplied the simulator uses default values that will, at least, provide a computable model, although probably not the model you want.

Level 1 is simple, like the JFET, where the increase in gate-junction bias attracts charge to form the channel and modulates its resistance to drain current. It is known as a "square law" device because of the following expression relating drain current to gate-junction voltage (see Figure 16-18):

$$I_{drain} = \left(\frac{width}{length} \cdot \frac{\text{KP}}{2} \right) \cdot \left(V_{GS} - V_{threshold} \right)^2 \tag{16-26}$$

Unlike the JFET, the transconductance parameter, KP, relates device size to drain current. As you may recall, with the JFET it was necessary to approximate the gate capacitance as a linear function of the gate-junction voltage (which in turn describes how the channel region is modulated). For the MOS transistor, the capacitance is

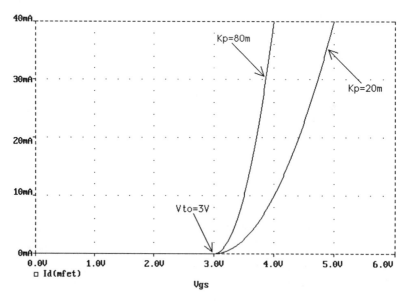

FIGURE 16-18 N-channel MOSFET drain current.

set, substantially, by the thickness of the gate oxide and the area of the gate, neither of which varies. This forms a linear, and nearly perfect capacitor; the same materials are used in some memory devices that store charge for a useful life measured in decades.

The square law result applies only when V_{DS} is greater than $V_{GS}-V_{threshold}$ (the *pinch-off* voltage), when the channel of the MOSFET is *saturated*. When V_{DS} is below pinch-off, the expression relating drain current to gate-junction voltage is

$$I_{drain} = \left(\frac{width}{length} \cdot \frac{KP}{2} \right) \cdot \left(2 \cdot \left(V_{GS} - V_{threshold} \right) \cdot V_{DS} - V_{DS}^{2} \right) \tag{16-27}$$

which describes (on an I_D versus V_{DS} plot) an inverted parabolic curve passing through the origin and, at its peak value (when V_{DS} equals the pinch-off voltage), intersects the square law formula. This parabolic region of operation is called the *linear* region; for small drain voltages, the expansion of the equation (above) is dominated by the linear term

$$I_{drain} \approx \left(\frac{width}{length} \cdot KP \right) \cdot \left(V_{GS} - V_{threshold} \right) \cdot V_{DS} \tag{16-28}$$

Finally, I_{drain} is zero when V_{GS} is less than $V_{threshold}$ (see Figure 16-19).

Nonlinear capacitance models are available for the MOSFET regardless of model level. Semiconductor *p-n* junction capacitance between the substrate (bulk) and source, or drain, is modeled the same way as for the diode (except that diffusion capacitance is not included, as these junctions are normally reversed-biased). Overlap capacitance, which is the excess overlap of the gate over any of the other sections of

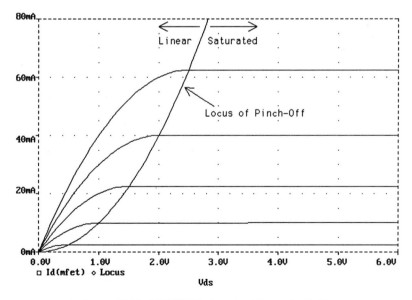

FIGURE 16-19 MOSFET drain current "curve family."

the device (due to the manufacturing process), is modeled as a fixed, stray capacitance to be added to any other calculated values. Overlap capacitance may be specified for any level of the MOS model. The remaining capacitance to be calculated is due to the operation of the intrinsic MOSFET; that is, these capacitance values come from the electrical characteristics of the charges in the channel and not the physical implementation of the device.

There are two models for the channel charge related capacitances:

- the Meyer model, which empirically splits the total capacitance into varying amounts between the gate and any other terminal, and

- the Ward-Dutton model (available for model levels 2 and 4), which calculates the distribution of charge and uses a three-terminal, nonreciprocal capacitor model.

For the level 2 model, the XQC parameter selects which model is used: if XQC has a value greater than 0.5 the Meyer model is used; otherwise the Ward-Dutton model is used. **Both models conserve charge.** The charge conservation rumors are due to a bad reputation acquired by an earlier version of SPICE2. Charge conservation is a problem of numerical integration and independent of the theory of the underlying capacitance models.

16.7 NONLINEAR MAGNETICS

MicroSim developed a nonlinear magnetics device based on the Jiles-Atherton magnetics model. This model is based on existing ideas of domain wall motion, including flexing and translation, to simulate the behavior of the magnetic material and thereby generate B-H curves. The slope of the B-H curves, then, set the inductance and current values for the windings associated with the magnetic core. The model accounts for the following nonlinear effects: initial permeability, saturation of magnetization, hysteresis (including coercivity and remanence), and dynamic core losses.

The Jiles-Atherton model supposes that the magnetic material is made up of loosely coupled domains that have an equilibrium B-H curve, called the *anhysteric*. This curve is the locus of B-H values generated by superimposing a DC magnetic bias and a large AC signal that decays to zero. It is the curve representing minimum energy for the domains and is modeled, in theory, by

$$M_{anhysteric} = M_{saturation} \cdot F\left(\frac{H_{effective}}{A}\right)$$

$$\text{where: } H_{effective} = H + \alpha \cdot M \tag{16-29}$$

$$F(x) = \frac{1}{tanh(x)} - \frac{1}{x}$$

For a given H (magnetizing influence), the anhysteric magnetization is the global flux level that the material would attain if the domain walls could move freely. Instead, the walls are stopped, or pinned, on dislocations in the material. The wall remains pinned until enough magnetic potential is available to break free and travel to the next pinning site. The theory supposes a mean energy required, per volume, to move domain walls. This is analogous to mechanical drag. A (simplified) equation of this is

$$\text{change in magnetization} = \frac{\text{potential}}{\text{drag}}$$

or

$$\frac{\partial M}{\partial H} = \frac{M_{anhysteric} - M}{K} \tag{16-30}$$

where K is the pinning-energy per volume (drag).

So much for irreversible domain wall motion. Reversible wall motion comes from flexing in the domain wall, especially when it is pinned at a dislocation, due to the magnetic potential (that is, the magnetization is not the anhysteric value). The theory supposes spherical flexure to calculate energy values and arrives at the (simplified) equation

$$\frac{\partial M}{\partial H} = C \cdot \frac{\partial \left(M_{anhysteric} - M \right)}{\partial H} \tag{16-31}$$

which must be added to the previous state equation. In Figure 16-20 you will see both the major B-H loop (where the magnetization is brought near to the positive and negative saturation value of the material) and some minor loops (where the magnetization is varied about an offset value). The locus of B versus H depends on

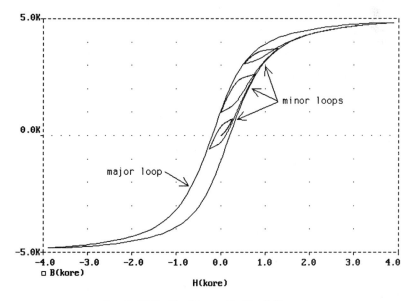

FIGURE 16-20 Nonlinear magnetics B-H curve.

the history of the material and does not follow the same path the way, say, a diode's DC current follows its DC bias voltage. This is one of the reasons that make magnetic materials difficult to model: there is not a single, explicit equation for B versus H.

Magnetic materials are used in inductor and transformer cores to provide high values of inductance in a small volume and to "trap" the majority of the magnetic flux within the windings for efficient energy transfer. Unfortunately, the materials are also nonlinear, which means that, since the value of $\partial B/\partial H$ (the slope of the curve) is proportional to the inductance of the component using the material as its core, the inductance therefore varies with the current through the windings. This effect can be seen more vividly by directly displaying, with Probe, the slope of the major loop. Notice how the trace starts at a low value, at H = 0; this value is proportional to the initial inductance of the material (see Figure 16-21).

Air gaps are available in the model. If the gap thickness is small compared with the other dimensions of the core, we can assume that all of the magnetic flux lines will go through the gap directly and that there will be little "fringing flux" (having a modest amount of fringing flux will only increase the effective air-gap length). In checking the field values around the entire magnetic path, we arrive at the equation

$$H_{core} \cdot L_{core} + H_{gap} \cdot L_{gap} = \sum n \times I$$

where the right-hand side is the sum of the amp-turns of the windings on the core. Also, we know that the magnetization in the air gap is negligible so that B_{gap} is equivalent to H_{gap}, making $B_{gap} = B_{core}$. These combine in the previous equation to yield

$$H_{core} \cdot L_{core} + B_{core} \cdot L_{gap} = \sum n \times I$$

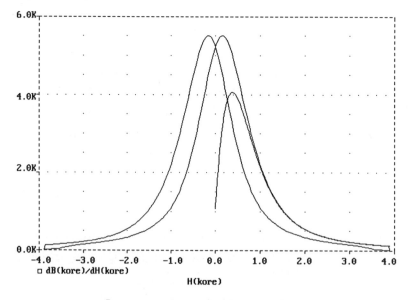

FIGURE 16-21 Plot of dB/dH (permeability).

This is a difficult equation to solve, especially for the Jiles-Atherton model, which is a state equation model rather than an explicit function (which you would expect, since the B-H curve depends on the history of the material). However, there is a graphical technique that solves for B_{core} and H_{core}, given $n \times I$, which is to

- take the nongapped B-H curve,

- extend a line from the current value of $n \times I$ (on the H-axis) with a slope of $-L_{core}/L_{gap}$ (this would be vertical if $L_{gap} = 0$), then

- find the intersection of the line with the B-H curve.

The intersection point is the value for B_{core} and H_{core} for the $n \times I$ of the gapped core. The $n \times I$ value is the apparent, or external, value of H_{core}, but the real value of H_{core} is less. This results in a smaller value for B_{core} and the "sheared over" B-H curves of a gapped core. PSpice implements the numerical equivalent of this graph technique.

The resulting B and H values are recorded in the Probe data file as B_{core} and $H_{apparent}$, since these are what the circuit "sees."

16.8 REFERENCES

For SPICE users who want to know more about the models they are using and how they relate to the component operation, I suggest the following texts:

P. ANTOGNETTI and G. MASSOBRIO, *Semiconductor Device Modeling with SPICE*, McGraw-Hill Company. This reference is a very comprehensive guide to the physics, and the derivation of the equations, for the devices in SPICE2. If you can buy only one of the items in this list, get this one.

DAVID A. HODGES and HORACE G. JACKSON, *Analysis and Design of Digital Integrated Circuits*, McGraw-Hill Company. This book has excellent sections on diode, bipolar transistor, MOSFET, and (in the latest edition) GAASFET models and how they relate to digital IC design. Examples and exercises are given for both hand calculations and SPICE simulation.

IAN GETREU, *Modeling the Bipolar Transistor*, Tektronix, Inc., part # 062-2841-00. This is the standard reference for a concentrated look at the development of bipolar transistor models. The only shortcoming is that the book was printed in 1976, before SPICE2 was released, so the model development does not include the extensions of the SPICE2 bipolar transistor.

ANDREW S. GROVE, *Physics and Technology of Semiconductor Devices*, and
S. M. SZE, *Physics of Semiconductor Devices*, both from John Wiley & Sons, Inc. These books have appropriate titles: more physics are included in these books, and they are the standard semiconductor physics references. These texts focus on device operation and only mention circuitry and uses in passing. While there

are no references to circuit simulation, all the derivations for the formulas SPICE uses, like the Shockley equation, are included.

D. C. JILES and D. L. ATHERTON, "Theory of ferromagnetic hysteresis," *Journal of Magnetism and Magnetic Materials*, 61, 48 (1986). This is a good reference for those who want the physics behind the magnetics model in PSpice.

ANDREI VLADIMIRESCU and SALLY LIU, "The Simulation of MOS Integrated Circuits Using SPICE2," Memorandum No. M80/7 (February 1980). This document describes, in detail, the MOS levels 2 and 3 device equations. Get this only if you are serious, as the text is fairly terse and knowledge of MOS device physics is assumed. This document is available by sending a check for $10 payable to *The Regents of the University of California* to this address:

> Cindy Manly-Fields
> EECS/ERL Industrial Liaison Program
> 497 Cory Hall
> University of California
> Berkeley, California 94720

B. J. SHEU, D. L. SCHARFETER, P. KO, and M. JENG, "BSIM: Berkeley Short-Channel IGFET Model for MOS Transistors," *IEEE Journal of Solid-State Circuits*, vol. 22, no. 4, pp. 558-66 (1987). This paper describes the "level 4" MOS model in SPICE2 and SPICE3. This model is reported to work well for small devices and has a parameter set that may be automatically extracted using semiconductor measurement equipment.

AN EXAMPLE CIRCUIT

Most technical books show examples to illustrate "typical use" of equipment or a program, and many books on circuit simulation follow this plan. We all want to see an example of our circuit *du jour*, so we can use our time efficiently (that is, cheat), run a few simulations, and move on to the next task. But being "spoon fed" working examples has limited long-term value for the student. We wouldn't be interested in circuit simulation if we could always find exactly the right circuit. Or, if we could do the design ourselves, correctly, why bother verifying the function with a simulation?

The time-tested benefit of circuit simulation is to reveal problems. Occasionally, even the best engineer will overlook some aspect of a design. Circuit simulation is one way, of many ways, to verify correct operation. The computer is not blind-sighted by the intended function of the circuit, as its creator might be. Then, through good judgement and experimentation, the circuit is improved to reduce the number of problems (or discarded); the simulator becomes a compass to help you "navigate" toward the final design. Just keep in mind that, like making breadboard tests, when a simulation result looks correct this may only mean that you still have not seen the symptoms of remaining problems.

So now, to show a practical use of circuit simulation, we will examine a circuit that **does not** work as expected. Then, we will discover what is wrong and **understand** what improvements could be made to the circuit.

THE EXAMPLE CIRCUIT

What could be simpler than a single stage, common emitter, AC-coupled amplifier, like the one shown in Figure A-1? The design should have a gain slightly less than five, from the ratio *RC/RE*. With small-signal analysis we should expect even gain

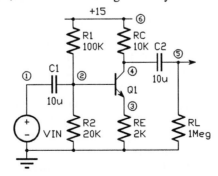

FIGURE A-1   The example circuit.

through the audio frequencies, but there will be a surprise at high frequencies. Then, we will find trouble in the transient analysis. But first, we have our circuit file:

```
AC-coupled amplifier
Vpwr 6 0 DC 15V
Vin 1 0 AC 1 SIN(0 2V 1KHz)
C1 1 2 10u
R1 6 2 100K
R2 2 0 20K
RC 6 4 10K
Q1 4 2 3 bjt
RE 3 0 2K
C2 4 5 10u
RL 5 0 1Meg
.model bjt npn(bf=80 cjc=5p rb=100)
.probe
.ac dec 5 10m 100Meg
.tran .02ms 2ms 0 .01ms
.end
```

The file starts with a title line. Next, we have described a power source and an input stimulus. The input stimulus includes both a small-signal and a transient specification. As noted earlier in this text, it is simpler to use an AC source with a magnitude of one as all of the responses can be interpreted as gain or attenuation.

The circuit devices are described next, followed by a simple model for the transistor. A Probe graphics data file is requested, for simulation results. Then the actual analyses are specified. The small-signal analysis will extend from 0.01Hz to 100MHz. The transient analysis runs for two cycles at 1KHz, and uses a step-ceiling to produce ultra-smooth graphic results.

The small-signal results look promising, as shown in Figure A-2. The mid-band gain is just shy of five (about 14dB) and the audio frequencies are covered. A surprise shows up at the highest frequencies: the gain does not fall below unity. A quick look back at the circuit and we realize that the input signal will feed forward through the transistor's collector-base ("Miller") capacitance. At these frequencies, the input drives the output directly and we might consider additional filtering if feed-forward will pose a problem in the system this amplifier will be in.

The transient results provide a shock, as shown in Figure A-3. The output, at node 5, is clipped and exhibits phase inversion in the trough of the waveform. How can this be?

Well, our input is a 2-volt (peak) sinewave, or 4 volts peak-to-peak. With a gain of five the output should be 20 volts peak-to-peak, an excursion that is greater than the voltage of the power supply to the amplifier. So the design is flawed in that the input needs to be something less than the 2-volt (peak) level to prevent clipping. Or, you may decide to increase the voltage of the power supply to provide the latitude needed for a 2-volt (peak) input... it depends on the situation.

It would be interesting to follow-up on the phase inversion in the trough of the output waveform. In an AC-coupled design, like this, you might not expect to see this kind of behavior, and yet there it is.

Since the amplifier's output is 180° out of phase from the input, at this frequency, the phase inversion is occurring not at the trough of the input signal, but at its crest.

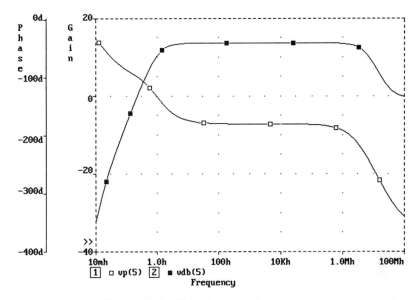

FIGURE A-2   Small-signal results: phase and gain.

This means that there is something happening as the base of the transistor is driven high by the input signal, and the collector of the transistor is being driven low by the amplification provided by transistor action and a resistive load. The previous plot of input and output signals is helpful, but these signals are AC-coupled to the "real"

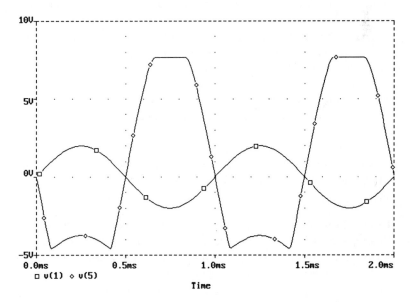

FIGURE A-3   Transient results show output distortion.

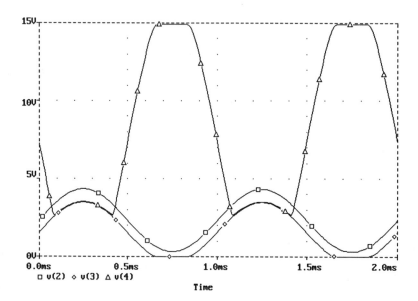

FIGURE A-4 Plot of transistor terminal voltages.

amplifier. It might be helpful to plot the terminal voltages of the transistor to inspect the "action" there. This is done in Figure A-4.

Now we can see what is happening! During the troughs of the input (crests of the output) the transistor shuts-off: the base-to-emitter voltage decreases, since the base voltage is driven by the input signal and the emitter voltage cannot go below ground potential. The collector voltage goes to the supply potential, as the collector current drops to zero.

During the crests of the input (troughs of the output) the transistor becomes saturated: the base-to-collector voltage, normally negative, becomes positive as the base voltage is driven by the input signal and the collector voltage is the result of transistor action and a resistive load. The collector potential cannot be less than the emitter potential. The output signal appears to have a phase inversion as the base-collector junction of the transistor, now forward biased, allows the input signal to drive the output directly! The collector voltage is a "diode drop" below the base voltage.

We conclude this example by reducing the input signal to one volt (peak) by changing the circuit file description for `Vin` to

```
Vin 1 0 AC 1 SIN(0 1V 1KHz)
```

Running this simulation and re-plotting the input and output signals, we see, as in Figure A-5, that the output signal is much less distorted. But the transistor still seems to be on the verge of saturation during the crests of the input signal. Perhaps we should lower the quiescent operating current of the transistor by increasing the value of `R1`.

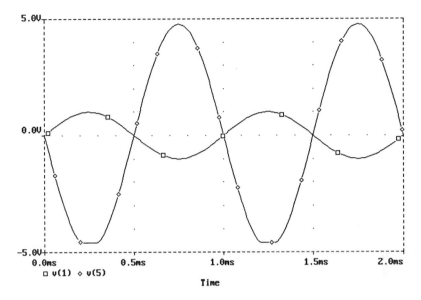

Figure A-5 Improved results.

We can make a quick calculation of a new value for R1 by looking back at Figure A-4. At time zero, the collector voltage is about 7 volts. This is a good starting point; but the collector voltage when the saturation occurs is about 3 volts, so the transistor only has $7 - 3 = 4$ volts of negative excursion available but has $15 - 7 = 8$ volts of positive excursion available. It would be better to balance the excursions by decreasing the operating current of the transistor so the quiescent collector voltage is closer to 9 volts. This means operating current will decrease in a ratio of 6:8, which is a 25% reduction in base current provided by R1. But we won't need to increase R1 by 25%, since this change will also decrease the quiescent base voltage and provide more room for negative excursions in the output signal before saturating the transistor. So, maybe only a 20% increase in R1 is required.

Of course, a few more simulations will verify the design changes. There may be other ways to improve the design for your needs, such as: increasing the power supply voltage, changing RC and RE to provide the same gain but a different operating current, etc. Or, you may discard this circuit and try a different one.

ABRIDGED SUMMARY OF PSPICE STATEMENTS

This section will quickly review control statements available in PSpice as of July, 1994. Each statement is described by an example of its use in the circuit file, with some comments. Often there are more detailed comments in the pages of this text. Also, you may want to refer to the *Circuit Analysis Reference Manual* (available from MicroSim Corporation).

---

*
Comment

General Form

> * *any text*

Example

```
* This is an example of a comment
```

A statement beginning with "*" is a comment line and has no effect. The use of comment statements throughout your circuit file is recommended. It is good practice to place a comment just before a subcircuit definition to identify the nodes; for example:

```
*              +IN -IN V+ V- +OUT -OUT
.SUBCKT OPAMP 100 101 1   2   200   201
```

or to identify major blocks of circuitry.

;
In-Line Comment

General Form

> *circuit file text* ; *any text*

Examples

```
R13  6  8  10K  ;feedback resistor
C3  15  0  .1U  ;decouple supply
```

A ";" is treated as the end of a line: PSpice moves on to the next line in the circuit file. The text after the ";" is a comment and has no effect. The use of comments throughout the input is recommended. This type of comment can also replace comment lines, which must start with "*" in the first column.

.AC   Small-Signal Analysis

General Form

.AC {LIN|DEC|OCT} *points start_frequency end_frequency*

Examples

```
.AC LIN 101 100Hz   200Hz
.AC DEC  20  1MEG 100MEG
.AC OCT  10  1kHz  16kHz
```

The .AC statement is used to calculate the small-signal response of a circuit at specific frequencies over a range. LIN, DEC, or OCT specify the sweep type and *points* is the number of points in the sweep.

LIN   Linear sweep. The analysis frequency is swept linearly from *start_frequency* to *end_frequency*, and *points* is the **total** number of points in the sweep.

DEC   Log sweep, by decades. The analysis frequency is swept logarithmically, by decades, from *start_frequency* to *end_frequency*, and *points* is the number of points per decade.

OCT   Log sweep, by octaves. The analysis frequency is swept logarithmically, by octaves, from *start_frequency* to *end_frequency*, and *points* is the number of points per octave.

.DC   DC Sweep

General Forms

.DC [[LIN]] *sweep_variable start end increment* [[*nested_sweep*]]

.DC {DEC|OCT} *sweep_variable start end points* [[*nested_sweep*]]

.DC *sweep_variable* LIST *value...* [[*nested_sweep*]]

Examples

```
.DC VIN -.25 .25 .05
.DC LIN I2 5mA -2mA 0.1mA
.DC VCE 0V 10V .5V  IB 0mA 1mA 50uA
.DC RES RMOD(R) 0.9 1.1 .001
.DC DEC NPN QFAST(IS) 1E-18 1E-14 5
.DC TEMP LIST 0 20 27 50 80 100
.DC PARAM Vsupply 7.5 15 .5
```

The .DC statement causes a DC sweep to be performed on the circuit. The DC sweep analysis calculates a circuit's bias point over a range of values for *sweep_variable*. See Chapter 4 for more on the use of this type of analysis.

The *nested_sweep* is optional. A second *sweep_variable*, type of sweep, *start*, *end*, and *increment* (or *points*) may be placed after the first sweep. In this case the first sweep will be the "inner" loop: the entire first sweep will be done for each value of the second sweep. The rules for the values in the second sweep are the same as for the first. The second sweep generates an entire .PRINT table or .PLOT plot for each value of the sweep. Probe displays nested sweeps as a family of curves.

After the DC sweep is finished *sweep_variable* is set back to the value it had before the sweep started.

The sweep can be linear, logarithmic, or a list of values. The keyword LIN is optional for linear sweeps. The sweep types are

LIN   Linear sweep. The *sweep_variable* is swept linearly from the value of *start* to *end*, and *increment* is the step size.

DEC   Log sweep, by decades. The *sweep_variable* is swept logarithmically, by decades, from the value of *start* to *end*, and *points* is the number of steps per decade.

OCT   Log sweep, by octaves. The *sweep_variable* is swept logarithmically, by octaves, from the value of *start* to *end*, and *points* is the number of steps per octave.

LIST   Use a list of values. In this case there are no start and/or end values. Instead, the numbers that follow the keyword LIST are the values to which *sweep_variable* will be set in sequence.

The *sweep_variable* can be one of the following types:

Source: a name of an independent voltage, or current, source. During the sweep the source's voltage or current is set to the sweep value.

Model parameter: *model_type   model_name* (*parameter_name*)
The parameter in the model is set to the sweep value. Note that model temperature parameters, such as TC1 and TC2 for the resistor, cannot be usefully swept as temperature updates are processed before the .DC sweep.

Temperature: use the keyword TEMP for *sweep_variable*. The temperature is set to the sweep value. For each value in the sweep, all the circuit components have their model parameters updated to that temperature.

Global parameter: use the keyword PARAM followed by the parameter name for *sweep_variable*. During the sweep, the global parameter's value is set to the sweep value and all expressions are reevaluated.

.DISTRIBUTION — User-Defined Distribution

General Form

> .DISTRIBUTION *name* (*deviation*, *probability*)...

Examples

> .DISTRIBUTION bi_modal (-1,1) (-.5,1) (-.5,0) (.5,0) (.5,1) (1,1)
> .DISTRIBUTION triangular (-1,0) (0,1) (1,0)

The .DISTRIBUTION statement is used to define a distribution of tolerances for use with Monte Carlo analysis **only**. The curve described by a .DISTRIBUTION statement controls the relative probability distribution of random numbers generated by PSpice to calculate model parameter deviations. Several distributions can be defined, and each is referenced by "name" when used to specify the tolerance distribution of a model parameter (see the .MODEL statement). A distribution defined this statement can be selected as the default by using the DISTRIBUTION parameter in the .OPTIONS statement.

The distribution curve is defined by deviation, probability pairs (as corner points) in a piecewise-linear fashion. Up to 100 such pairs are allowed. Each *deviation* must be in the range $(-1, +1)$, which matches the range of the random number generator. No *deviation* may be less than the previous *deviation* in the list, although it may repeat the previous value. Each *probability* represents a relative probability, and must be positive or zero.

.END — End of Circuit

General Form

> .END

Example

> .END

The .END statement marks the end of a circuit. All data and commands must come before it. When the .END statement is reached, PSpice does all the specified analyses on the circuit. The last statement in an input file **should** be a .END statement.

There may be more than one circuit in an input file. Each circuit and its commands are marked by a .END statement. PSpice processes all the analyses for each circuit before going on to the next one. Everything is reset at the beginning of each circuit. Having several circuits in one file gives the same results as having them in separate files and running each one separately. This is a convenient way to arrange a set of runs to be done unattended.

.ENDS   End of Subcircuit Definition

General Form

> .ENDS [[*subcircuit_name*]]

Examples

> .ENDS
> .ENDS OPAMP

The .ENDS statement marks the end of a subcircuit definition (started by a .SUBCKT statement). It is good practice to repeat the subcircuit name, although this is not required.

.EXTERNAL   External Port

General Form

> .EXTERNAL {INPUT|OUTPUT|BIDIRECTIONAL} *node...*

Examples

> .EXTERNAL INPUT Data1, Data2
> .EXTERNAL OUTPUT TP1
> .EXTERNAL BIDIRECTIONAL BiPort1, BiPort2

The .EXTERNAL statement identifies "external port" nets for use by PSpice A/D and PLogic digital simulations. Please refer to MicroSim's *Circuit Analysis Reference Manual* for details regarding this statement.

.FOUR   Fourier Analysis

General Form

> .FOUR *frequency* [[*last_harmonic_value*]] *output_value...*

Example

> .FOUR 10kHz V(5) V(6,7) I(VSENS3)
> .FOUR 120Hz 17 V(20)

Fourier analysis performs a decomposition into Fourier components of the result(s) of a transient analysis. A .FOUR statement requires a .TRAN statement. By default the first nine harmonics are analyzed, but the optional [[*last_harmonic_value*]] will set a new limit. See Chapter 12 for more on the use of this type of analysis.

.FUNC   Function Definition

General Form

.FUNC *name* ([[*arg...*]])   { *body* }

Examples

```
.FUNC E(x)        {exp(x)}
.FUNC HypSin(x)   {(E(x)-E(-x))/2}
.FUNC MIN(A,B)    {(A+B-ABS(A-B))/2}
.FUNC MAX(A,B)    {(A+B+ABS(A-B))/2}
.FUNC F()         {1/(6.28*sqrt(L*C))}
```

The .FUNC statement is used to define "functions" that may be used in expressions. Besides their obvious flexibility, they are also useful where there are several similar subexpressions in a circuit file.

Functions **may not** be redefined. Function names **must not** be the same as built-in functions described in §2.4 (page 10): for example, "sin."

The *body* of a definition may refer to other (previously defined) functions: the second example, HypSin(x), uses the first example, E(x).

Up to 10 arguments may be used in a definition. The number of arguments in the use of a function must agree with the number in the definition. Functions may be defined with no arguments, but the parentheses are still required. Parameters, as shown in the last example, and the Laplace variable "s" are allowed in function definitions.

The *body* of a defined function should be enclosed in curly braces "{ }" as shown in the examples. Previous versions of PSpice did not require this and a warning may be issued regarding compatibility.

Hint: you may want to create a file of popular .FUNC definitions and access them with a .INC statement near the beginning of a circuit file.

.IC   Initial Bias-Point Condition

General Form

.IC V(*node*[[,*node*]]) =*value...*

Examples

```
.IC V(2)=3.4  V(102)=0  V(3)=-1V
.IC V(InPlus,InMinus)=1e-3  V(100,133)=5.0V
```

The .IC statement is used to set initial conditions for both small-signal and transient bias points. Initial conditions may be given for any, or all, of the circuit's nodes. Also, the voltage between two nodes may be specified. Each *value* is a voltage that is assigned to *node* for the duration of the bias-point calculation. After the bias point has been calculated and the analysis started, the node is "released."

The .IC statement sets the initial conditions for the bias point **only**, and **does not** affect the DC sweep. If the circuit contains both .IC and .NODESET statements, the .NODESET statements are **ignored**.

.INC   Include File

General Form

> .INC *"file_name"*

Examples

> .INC "SETUP.CIR"
> .INC "C:\LIB\VCO.CIR"

The .INC statement is used to insert the contents of another file. The *file_name* can be any character string that is a legal file name for your computer system. It may include a volume, directory, and version number. On some systems, such as the VAX, it can also be a logical name.

Included files may contain any statements except: no title line is allowed (use a comment), and a .END statement (if present) marks only the end of the included file. An include file may contain a .INC statement (up to four levels of "including").

Including a file is the same as simply bringing that file's text into the circuit file. Everything in the included file is read in, every model and subcircuit definition, even if not needed.

.LIB   Library File

General Form

> .LIB *"file_name"*

Examples

> .LIB
> .LIB "LINEAR.LIB"
> .LIB "C:\LIB\BIPOLAR.LIB"

The .LIB statement is used to reference a model or subcircuit library in another file. The *file_name* can be any character string that is a legal file name for your computer system. It may include a volume, directory, and version number. On some systems, such as the VAX, it can also be a logical name. The file extension **is not** defaulted to ".LIB". If you specify a file name, you **must** include its extension.

Library files may contain comments, .MODEL statements, subcircuit definitions (including the .SUBCKT and .ENDS statements), and .LIB statements. No other statements are allowed.

If *file_name* is left off, all references will be done to the master library file, NOM.LIB, which, then, references the individual library files. When a library file is referenced, PSpice will first search for the file in the current working directory, and then in the directory specified by the environment variable PSPICELIB.

When any library is modified, PSpice creates an index file the first time the library is used. Thereafter, the index file is used to accelerate the look-up of library references. The index file is organized in a way that allows PSpice to locate a particular .MODEL or .SUBCKT quickly, in spite of how large the library file is.

.LOADBIAS   Load Bias-Point File

General Form

```
.LOADBIAS "file_name"
```

Examples

```
.LOADBIAS "SAVETRAN.NOD"
.LOADBIAS "C:\PROJECT\INIT.FIL"
```

The .LOADBIAS statement is used to load the contents of a bias-point file. Normally, the bias-point file will have been produced by a previous circuit simulation using the .SAVEBIAS statement. The *file_name* can be any character string that is a legal file name for your computer system. It may include a volume, directory, and version number. On some systems, such as the VAX, it can also be a logical name.

The bias-point file is a text file that contains one or more comment lines, and a .NODESET statement with the bias-point voltage values. If you want to set a fixed value for a transient analysis bias point, you can edit the bias-point file and replace the .NODESET statement with a .IC statement. Any nodes mentioned in the loaded file that are not present in the circuit, will be ignored after producing a warning message.

To echo the .LOADBIAS file contents to the output file, use the EXPAND option on the .OPTIONS statement.

.MC   Monte Carlo Analysis

General Form

```
.MC #runs {AC|DC|TRAN} output_value function [[option...]] [[SEED=value]]
```

Examples

```
.MC 10 TRAN V(5) YMAX
.MC 50 DC IC(Q7) YMAX LIST
.MC 20 AC VP(13,5) YMAX LIST OUTPUT(ALL)
.MC 10 TRAN V(3) YMAX SEED=9321
```

The .MC statement causes a Monte Carlo (statistical) analysis of the circuit. Multiple runs of the selected analysis (AC, DC, transient) are done. The first run is done with nominal values of all components, and all analyses that the circuit contains are performed during this "nominal" pass. Subsequent runs are done only for the selected analysis. Each subsequent run uses variations on model parameters as specified by the DEV and LOT tolerances on each .MODEL parameter (see the .MODEL statement for details on DEV and LOT tolerances). The *#runs* is the total number of runs to do. The *output_value* has the same format as for specifying .PRINT output.

The other specifications on the .MC statement control the kind and amount of output generated by the Monte Carlo analysis. The *function* specifies the operation to be performed on the results from *output_value* to reduce this vector (a waveform of, for example, a voltage vs. time or a current vs. frequency, etc.) to a single value. This value is the basis for the comparisons between the nominal and subsequent runs. The *function* must be one of the following:

YMAX will find the **greatest difference** in each waveform from the nominal run.

MAX will find the **maximum value** of each waveform.

MIN will find the **minimum value** of each waveform.

RISE_EDGE(*value*) will find the **first occurrence** of the waveform crossing **above** the threshold *value*. The waveform must have one or more points at or below *value* followed by one above; the output value listed will be where the waveform rises above *value*.

FALL_EDGE(*value*) will find the **first occurrence** of the waveform crossing **below** the threshold *value*. The waveform must have one or more points at or above *value* followed by one below; the output value listed will be where the waveform falls below *value*.

The [[*option...*]] includes zero or more of the following:

LIST will print out, at the beginning of each run, the model parameter values used for each component during that run.

OUTPUT(*type*) requests output from subsequent runs, after the nominal (first) run. The contents of the output from these runs is governed by the .PRINT, .PLOT, and .PROBE statements in the file. If OUTPUT(*type*) is omitted, then only the nominal run produces output. The *type* is one of the following:

ALL generates output for all runs, including the nominal run.

FIRST *n_runs* generates output only for the first *n_runs* runs.

EVERY *nth_run* generates output for every *nth_run* run.

RUNS *run_n...* does the analysis and generates output only for the listed runs. Up to 25 values may be specified in the list.

RANGE(*low*,*high*) restricts the range over which *function* will be evaluated. A "*" can be used to indicate "for all values." For example:

YMAX RANGE(*,.5) will evaluate YMAX for the waveform for values of the sweep variable (time, frequency, etc.) of .5 or less.

MAX RANGE(-1,*) will find the maximum of the waveform for values of the sweep variable (time, frequency, etc.) of −1 or more.

If RANGE(*low*, *high*) is omitted, then *function* is evaluated over the whole sweep range. This is equivalent to RANGE(*, *).

[[SEED=*value*]] defines the seed value for the random number generator within the Monte Carlo analysis (see *The Art of Computer Programming*, Donald Knuth, vol. 2, p. 171, "subtractive method"). The *value* must be an odd integer ranging from 1 to 32,767. If the seed value is not explicitly set, it will default to 17,533. For normal use, you will want to use the default seed value in order to achieve a constant set of results from several simulations. Changing the seed value will force a different sequence of random variations in the model parameters.

.MODEL   Model

General Form

.MODEL *name* [[AKO:*reference*]] *type* ([[*parameter*...]])

where *parameter* has the form: *name* = *value* [[*tolerance*]]

Examples

```
.MODEL RMAX    RES (R=1.5 TC1=.02 TC2=.005)
.MODEL DNOM    D   (IS=1E-9)
.MODEL QDRIV   NPN (IS=1E-7 BF=30)
.MODEL MLOAD   NMOS(LEVEL=1 VTO=.7 CJ=.02pF)
.MODEL CMOD    CAP (C=1 DEV 5%)
.MODEL DLOAD   D   (IS=1E-9 DEV .5% LOT 10%)
.MODEL RTRACK  RES (R=1 DEV/GAUSS 1% LOT/UNIFORM 5%)
.MODEL QDR2 AKO:QDRIV NPN (BF=50 IKF=50m)
```

The .MODEL statement defines a set of parameters that characterize a particular model. These can be referenced by devices in the circuit by *name*, which must start with a letter. It is good practice to make this the same letter as the device type in the circuit (for example, D for diode), but this is not required.

The last example uses the AKO: (A KIND OF) syntax to reference the parameters of the model QDRIV from the third example. The value of each parameter of the referenced model is used unless over-ridden by the current model, which must be the same model type.

The *type* is the device type and must be one of:

type	Instance name	Type of device
CAP	C*xxx*	capacitor
IND	L*xxx*	inductor
RES	R*xxx*	resistor
D	D*xxx*	diode
NPN	Q*xxx*	NPN bipolar transistor
PNP	Q*xxx*	PNP bipolar transistor
LPNP	Q*xxx*	lateral PNP bipolar transistor
NJF	J*xxx*	N-channel junction FET
PJF	J*xxx*	P-channel junction FET

type	Instance name	Type of device
NMOS	M*xxx*	N-channel MOSFET
PMOS	M*xxx*	P-channel MOSFET
GASFET	B*xxx*	N-channel GaAs MESFET
CORE	K*xxx*	nonlinear, magnetic core (transformer)
VSWITCH	S*xxx*	voltage-controlled switch
ISWITCH	W*xxx*	current-controlled switch
DINPUT	N*xxx*	digital input device (receive from digital)
DOUTPUT	O*xxx*	digital output device (transmit to digital)
UIO	U*xxx*	digital IO model
UGATE	U*xxx*	standard gate
UTGATE	U*xxx*	tri-state gate
UBTG	U*xxx*	bidirectional transfer gate
UEFF	U*xxx*	edge-triggered flip-flop
UGFF	U*xxx*	gated flip-flop
UDLY	U*xxx*	digital delay line
UPLD	U*xxx*	programmable logic array
UROM	U*xxx*	read-only memory
URAM	U*xxx*	random access (read/write) memory
UADC	U*xxx*	multi-bit analog-to-digital converter
UDAC	U*xxx*	multi-bit digital-to-analog converter

Devices can reference models only of the correct type. A JFET can reference a model of types NJF, or PJF, but not of type NPN. There can be more than one model of the same type in a circuit, although they must have different names.

Following *type* is a list of parameter values enclosed by parenthesis. None, any, or all parameters may be assigned values. Default values are used for all unassigned parameters. The lists of parameter names, meanings, and default values are located with the individual device descriptions.

The *tolerance_spec* may be appended to each parameter, with the format

$$[\text{DEV } [\![track\_\&\_dist]\!] \ value \ [\![\%]\!]] \quad [\text{LOT } [\![track\_\&\_dist]\!] \ value \ [\![\%]\!]]$$

to specify individual device (DEV) and device lot (LOT) parameter value deviations. The *tolerance_spec* is used by the Monte Carlo analysis only. DEV tolerances are independent; LOT tolerances track: all devices that refer to the same model will use the same value of the model parameter. The optional [[%]] indicates a relative (percentage) tolerance; if it is omitted *value* is in the same units as the parameter itself.

The optional [[*track_&_dist*]] specifies the tracking lot and name of the distribution to use, with the format

$$[\![/gen\#]\!] \quad [\![/distribution]\!]$$

These specifications must immediately follow the keywords DEV and LOT, **without** spaces.

The $[\![/gen\#]\!]$ specifies which of ten random number generators, 0 through 9, are used to calculate parameter value deviations. This allows deviations to be correlated between parameters in the same model, as well as between models. The generators for DEV and LOT tolerances are distinct: there are ten generators for DEV tracking and ten generators for LOT tracking. Tolerances without $[\![/gen\#]\!]$ get individually generated random numbers.

The $[\![/distribution]\!]$ and the default distribution (set using the .OPTIONS statement DISTRIBUTION parameter) are one of the following:

UNIFORM   generates uniformly distributed deviations over the range ±*value*.

GAUSS   generates deviations with a Gaussian distribution over the range ±3σ, and *value* specifies the ±1σ deviation (this will generate deviations greater than ±*value*).

user_name   generates deviations using a user-defined distribution, and *value* specifies the ±1 deviation in the user definition (see the .DISTRIBUTION statement.

Some device models have two levels of temperature specification that can be customized for each model. The temperature at which the model parameters were measured at is specified by T_MEASURED, which overrides the value TNOM set by the .OPTIONS statement.

Also, the "current" device temperature can be set to modify the "current" circuit temperature, as set by the .TEMP or .STEP commands:

- T_ABS sets an absolute (fixed) temperature for the device

- T_REL_GLOBAL sets the device's temperature relative to the circuit's temperature, $T_{dev} = T_{ckt} +$ T_REL_GLOBAL

- T_REL_LOCAL sets the device's temperature relative to an AKO device's absolute temperature, $T_{dev} =$ T_ABS$_{AKO} +$ T_REL_LOCAL

.NODESET   Nodeset

General Form

.NODESET V(*node* [[,*node*]])=*value* ...

Examples

```
.NODESET V(2)=3.4 V(102)=0 V(3)=-1V
.NODESET V(InPlus,InMinus)=1e-3 V(100,133)=5.0V
```

The .NODESET statement helps calculate the bias point by providing an initial guess for some nodes. Some or all of the circuit's nodes may be given an initial guess. Also, the voltage between two nodes may be specified. .NODESET is effective for the bias point (both small-signal and transient bias points) and for the first step of the DC sweep. It has no effect during the rest of the DC sweep, or during the transient analysis itself.

Unlike the .IC statement, .NODESET provides only an initial guess for some node voltages. It does not clamp those nodes to the specified voltages; however, by providing an initial guess, .NODESET may be used to "break the tie" in, for instance, a flip-flop, and make it "come up" in a desired state.

If both .IC and .NODESET commands are present, the .NODESET commands are ignored for the bias-point calculations (.IC overrides .NODESET).

.NOISE   Noise Analysis

General Form

.NOISE V(*node* [[,*node*]]) *name* [[*interval*]]

Examples

```
.NOISE V(5)    VIN
.NOISE V(101) VSRC 20
.NOISE V(4,5) ISRC
```

The .NOISE statement causes a noise analysis of the circuit to be done. Noise analysis is done in conjunction with small-signal analysis and requires there to be a .AC statement.

V(*node* [[,*node*]]) is an output voltage. It has a form such as V(5), which is the voltage at an output node, or a form such as V(4,5), which is the output voltage across two nodes. The *name* is the name of an independent voltage or current source at which the equivalent input noise will be calculated; *name* is not itself a noise generator, but only a place at which to calculate the equivalent input noise.

The noise-generating devices in a circuit are the resistors and the semiconductor devices. For each frequency of the small-signal analysis, each noise generator's contribution is calculated and propagated to the output nodes. There, all the propagated noise values are RMS-summed. The gain from the input source to the output voltage is also computed, which with the total output noise an equivalent input noise is calculated. If

name is a voltage source, then the input noise units are volt/Hertz$^{\frac{1}{2}}$

name is a current source, then the input noise units are amp/Hertz$^{\frac{1}{2}}$

The output noise units are always volt/Hertz$^{\frac{1}{2}}$ as it is a voltage.

If the optional [[*interval*]] is present, it specifies the print interval. Every *n*th frequency, where *n* is the print interval, a detailed table is printed showing the individual contributions of all the circuit's noise generators to the total noise. These values are the noise amounts propagated to the output nodes, not the noise amounts at each generator. If [[*interval*]] is not present, then no detailed table is printed.

The detailed table is printed while the analysis is being done, and does not need a .PRINT or a .PLOT statement. The output noise and equivalent input noise may be output with a .PRINT statement or a .PLOT statements if desired. Noise analysis is the only analysis for which you have a choice about using the .PRINT and .PLOT statements.

.OP   **Bias Point**

General Form

> .OP

Example

> .OP

The .OP statement causes detailed information about the bias point to be printed. The bias point is calculated whether or not there is a .OP statement. Without a .OP statement the only information about the bias point that is output is a list of the node voltages.

With a .OP statement the currents and power dissipation of all the voltage sources are printed. Also the small signal (linearized) parameters of all the nonlinear controlled sources and all the semiconductor devices are output.

The .OP statement controls output for the regular bias point only. The .TRAN statement controls output for the transient analysis bias point.

.OPTIONS   **Options**

General Form

.OPTIONS *option* [*=value*] ...

Examples

```
.OPTIONS NOECHO NOMOD DEFL=12u DEFW=8u DEFAD=150p DEFAS=150p
.OPTIONS ACCT RELTOL=.01
.OPTIONS DISTRIBUTION=GAUSS
```

The .OPTIONS statement is used to set all the options, limits, and control parameters for the various analyses including the output width (see the .WIDTH statement, which is still supported).

The options are listed in any order. There are two kinds of options: those with values and those without. The options without values are flags of various kinds and simply listing the option name is sufficient. The .OPTIONS commands are cumulative: if there are two (or more) .OPTIONS statements, the effect is the same as if all the options were listed together in one .OPTIONS statement. If the same option is listed more than once, only the latest value is used.

The following table lists the flag options. Flag options are normally "off" and are turned "on" by being specified.

option	**Meaning**
ACCT	summary and accounting information is output at the end of all the analyses
EXPAND	lists devices created by subcircuit expansion and lists contents of the bias-point file (see .SAVEBIAS and .LOADBIAS)
LIBRARY	lists lines used from library files
LIST	lists summary of circuit elements (devices)
NOBIAS	suppresses the printing of the bias-point node voltages
NODE	lists summary of connections (node table)
NOECHO	suppresses listing of the input file
NOMOD	suppresses listing of model parameters and temperature updated values
NOOUTMSG	suppresses simulation error messages in the output file
NOPAGE	suppresses paging and the banner for each major section of output
NOPRBMSG	suppresses simulation error messages in the Probe data file
NOREUSE	suppresses the automatic saving and restoring of bias-point information between different temperatures, Monte Carlo runs, worst case runs, and parametric analyses (.STEP).
OPTS	lists values for all options

The table below lists the options with values and their default values:

option	Meaning	Units	Default
ABSTOL	best accuracy of currents	amp	1pA
CHGTOL	best accuracy of charges	coulomb	0.01pC
CPTIME	CPU time allowed for this run	second	0
DEFAD	MOSFET default drain area (AD)	meter^2	0
DEFAS	MOSFET default source area (AS)	meter^2	0
DEFL	MOSFET default length (L)	meter	100μm
DEFW	MOSFET default width (W)	meter	100μm
DIGFREQ	minimum digital time step is 1/DIGFREQ	Hertz	10GHz
DIGDRVF	minimum drive resistance (UIO type model, DRVH and DRVL parameter values)	ohm	2
DIGDRVZ	maximum drive resistance (UIO type model, DRVH and DRVL parameter values)	ohm	20K
DIGERRDEFAULT	default error limit for digital constraint devices		
DIGERRLIMIT	maximum digital error message limit		0
DIGINITSTATE	init. state for flip-flops & latches: 0=clear, 1=set, 2=X		2
DIGIOLVL	default digital IO level: 1-4; see UIO model		1
DIGMNTYMX	default delay selector: 1=min, 2=typical, 3=max		2
DIGMNTYSCALE	factor to derive minimum delays from typical delays		0.4
DIGOVRDRV	ratio of drive resistances required to allow one output to override another driving the same node		3
DIGTYMXSCALE	factor to derive maximum delays from typical delays		1.6
DISTRIBUTION	default distribution; see comment following this table		UNIFORM
GMIN	minimum conductance used for any branch	ohm^{-1}	1E−12
ITL1	DC and bias-point "blind" iteration limit		40
ITL2	DC and bias-point "educated guess" iteration limit		20
ITL4	iteration limit at any point in transient analysis		10
ITL5	iteration limit for all points in transient analysis		0
LIMPTS	maximum points allowed for any print table or plot		0
NUMDGT	number of digits output in print tables (maximum 8 useful digits)		4
PIVREL	relative magnitude required for pivot in matrix solution		1E−3
PIVTOL	absolute magnitude required for pivot in matrix solution		1E−13
RELTOL	relative accuracy of V's and I's		0.001
TNOM	default temperature (also the temperature at which model parameters are assumed to have been measured)	°C	27
VNTOL	best accuracy of voltages	volt	1μV
WIDTH	same as the statement .WIDTH OUT=*value*		80

The DISTRIBUTION option has a name as its value to set the default distribution for Monte Carlo deviations. The normal setting is for a UNIFORM distribution. The default distribution is used for all of the deviations throughout the Monte Carlo analyses, unless specifically overridden for a particular tolerance. The value for the default distribution can also be set to GAUSS or a user-defined distribution. If a user-defined distribution is selected, a .DISTRIBUTION statement must be included in the circuit file to define the user distribution for the tolerances.

.PARAM   **Parameter Definition**

General Form

.PARAM *name* = { *value* | {*expression*} }...

Examples

```
.PARAM VSUPPLY = 5V
.PARAM VCC = 12V, VEE = -12V
.PARAM BANDWIDTH = {100kHz/3}
.PARAM PI = 3.14159, TWO_PI = {2*3.14159}
.PARAM VNUM = {2*TWO_PI}
```

The keyword .PARAM is followed by a list of names with values, which must be either constants or expressions. Constants do not need to be bracketed by "{" and "}" like the *expression*. The *expression* must contain only constants or previously defined parameters. The .PARAM statement cannot be used inside a subcircuit definition.

There are several predefined parameters:

TEMP   temperature (*reserved, not available this release*)

VT   thermal voltage (*reserved, not available this release*)

GMIN   shunt conductance for semiconductor *p-n* junctions

The *name* cannot be one of: the predefined parameters (above), TIME, or one of the .TEXT statement names.

Once defined, a parameter can be used in place of most numeric values in the circuit description. For example:

All model parameters and values on .IC and .NODESET statements.

All device parameters, such as AREA, L, NRD, Z0. This includes IC= values on capacitors and inductors, but **not** the transmission-line parameters NL and F, and **not** the "in-line" temperature coefficients for the resistor (however, parameters can be used for the TC1 and TC2 resistor model parameters).

All independent voltage and current source (V and I device) parameters **except** for values defining a piece-wise linear (PWL) source.

Not the E, F, G, and H device polynomial coefficient values.

Parameters **cannot** be used in place of node numbers, nor can the values on analysis statements (`.AC`, `.DC`, `.TRAN`, etc.) be parameterized.

`.PLOT`   **Plot**

General Forms

> `.PLOT AC` *output_variable...*

> `.PLOT {DC|NOISE|TRAN}` *output_variable...* ⟦*lower_limit* , *upper_limit*⟧

Examples

```
.PLOT DC V(3) V(2,3) V(R1) I(VIN) I(R2) IB(Q13) VBE(Q13)
.PLOT AC VM(2) VP(2) VM(3,4) VG(5) VDB(5) IR(D4)
.PLOT NOISE INOISE ONOISE DB(INOISE) DB(ONOISE)
.PLOT TRAN V(3) V(2,3) (0,5V) ID(M2) I(VCC) (-50mA,50mA)
.PLOT TRAN D(QA) D(QB) V(3) V(2,3)
.PLOT TRAN V(3) V(R1) V([RESET])
```

The `.PLOT` statement allows results from AC, DC, noise, and transient analyses to be output as "line printer" plots. These plots are made by using characters to draw the plot; hence they will work with any kind of printer.

Following the analysis type is a list of the *output_variable*s and (possibly) y-axis limits. Up to 8 *output_variable*s are allowed on one `.PLOT` statement, however an analysis may have any number of `.PLOT` statements. See §4.2 (page 24).

`.PRINT`   **Print**

General Form

> `.PRINT`⟦`/DGTLCHG`⟧ `{AC|DC|NOISE|TRAN}` *output_variable...*

Examples

```
.PRINT DC V(3) V(2,3) V(R1) I(VIN) I(R2) IB(Q13) VBE(Q13)
.PRINT AC VM(2) VP(2) VM(3,4) VG(5) VDB(5) IR(6) II(7)
.PRINT NOISE INOISE ONOISE DB(INOISE) DB(ONOISE)
.PRINT TRAN V(3) V(2,3) ID(M2) I(VCC)
.PRINT TRAN D(QA) D(QB) V(3) V(2,3)
.PRINT/DGTLCHG TRAN QA QB RESET
.PRINT TRAN V(3) V(R1) V([RESET])
```

The `.PRINT` statement allows results from AC, DC, noise, and transient analyses to be output as tables ("print tables"). The values of the *output_variable*s are printed as a table with each column corresponding to one *output_variable*. The `.PRINT/DGTLCHG` form is for digital *output_variable*s only: values are printed for each *output_variable* whenever one of these changes.

Following the analysis type is a list of the *output_variable*s. There is no limit to the number of output variables; the printout is split up depending on the width of the data columns (set with the `NUMDGT` option) and the output width (set with the `WIDTH` option). An analysis may have any number of `.PRINT` statements. See §4.2 (page 24).

.PROBE   **Probe**

General Form

> .PROBE⟦/CSDF⟧ ⟦*output_variable...*⟧

Examples

```
.PROBE
.PROBE V(3) V(2,3) V(R1) VM(2) VP(2) I(VIN) I(R2) IB(Q13)
+       VBE(Q13) VDB(5)
.PROBE/CSDF
.PROBE V(3) V(R1) V([RESET])
.PROBE D(QBAR)
```

The .PROBE statement writes the results from AC, DC, and transient analyses to a data file. See §4.5 (page 30).

The first form (with no *output_variables*) writes all the node voltages and all the device currents to the data file. The second form writes only those output variables specified to the data file. (Note that unlike the .PRINT and .PLOT statements there is no analysis name before the output variables.) The third example creates a data file in a text-format using the COMMON SIMULATION DATA FORMAT (CSDF), not a binary format. This format is primarily used for transfers between different computer types. The fourth example illustrates how to specify a regular node, the voltage across a resistor, and a node that has a name rather than a number. The last example only writes the output at digital node QBAR to the data file.

.SAVEBIAS   **Save Bias Point to File**

General Form

> .SAVEBIAS "*file_name*" {DC|OP|TRAN} ⟦NOSUBCKT⟧
> + ⟦TIME=*value* ⟦REPEAT⟧⟧
> + ⟦TEMP=*value*⟧ ⟦STEP=*value*⟧ ⟦MCRUN=*value*⟧
> + ⟦DC=*value*⟧ ⟦DC1=*value*⟧ ⟦DC2=*value*⟧

Examples

```
.SAVEBIAS "OPPOINT" OP
.SAVEBIAS "TRANDATA.BSP" TRAN NOSUBCKT TIME=10u
.SAVEBIAS "SAVETRAN.BSP" TRAN TIME=5n REPEAT TEMP=50.0
.SAVEBIAS "DCBIAS.SAV" DC
.SAVEBIAS "SAVEDC.BSP" DC MCRUN=3 DC1=3.5 DC2=100
```

The .SAVEBIAS statement is used to save the bias-point node voltages, for the specified analysis, to *file_name*. The *file_name* can be any character string that is a legal file name for your computer system. It may include a volume, directory, and version number. On some systems, such as the VAX, it can also be a logical name.

A circuit file may contain a .SAVEBIAS statement for each of the three analysis types. If the simulation parameters do not match the keywords and values in the .SAVEBIAS statement, then no file is produced. When ⟦NOSUBCKT⟧ is used, the node voltages for subcircuits are not saved.

The [[TIME=*value* [[REPEAT]]]] form is used to define the transient analysis time at which the bias point is to be saved. If [[REPEAT]] is not used, then the next bias point greater than or equal to TIME=*value* is saved. If [[REPEAT]] is used, then TIME=*value* is the interval at which the bias point is saved. However, only the latest bias point is saved; any previous times are overwritten. The [[TIME=*value* [[REPEAT]]]] form can be used **only** with transient analysis.

The [[TEMP=*value*]] form defines the temperature at which the bias point is to be saved. The [[STEP=*value*]] form defines the step value at which the bias point is to be saved. The [[MCRUN=*value*]] form defines the number of the Monte Carlo or worst case analysis runs for which the bias point is to be saved.

The [[DC=*value*]], [[DC1=*value*]], and [[DC2=*value*]] forms are used to specify the DC sweep value at which the bias point is to be saved. The [[DC=*value*]] form should be used if there is only one sweep variable. If there are two sweep variables, then the [[DC1=*value*]] form is used to specify the first sweep value and the [[DC2=*value*]] form is used to specify the second sweep value.

The saved bias-point information is in the following format: one or more comment lines indicating the circuit name, title, date and time of run, analysis, temperature, etc.; and a single .NODESET statement containing the bias-point voltage values.

Only one bias point is saved to the file during any particular analysis. At the specified time, the bias-point information and the operating point data for the active devices and controlled sources are written to the output file. When the supplied specifications on the command line match the "state" of the simulator during execution, the bias point is written out:

- For the first example, the small-signal operating point (.AC or .OP) bias point is saved.

- In the second example, the transient bias point is written out at the time closest to, but not less than 10µs. No bias-point information for subcircuits is saved.

- Use of the REPEAT keyword in the third example causes the bias point to be written out every 5ns when the temperature of the run is 50°C. Only one set (time) of bias-point information is saved in the file at any time, over-writing any "older" set. The REPEAT keyword may **only** be used with the transient analysis. To repetitively save the most recent bias point, use TIME=0 and REPEAT.

- In the fourth example, because there are no parameters supplied, only the very first DC bias point is written to the file.

- The fifth example saves the DC bias point when the following three conditions are all met: the first DC sweep value is 3.5, the second DC sweep value is 100, and the simulation is on the third Monte Carlo run. If only one DC sweep is being performed, then the keyword DC can be substituted for DC1.

.SENS   Sensitivity Analysis

General Form

.SENS *output_variable...*

Example

.SENS V(9) V(4,3) V(17) I(VCC)

The .SENS statement causes a DC sensitivity analysis to be performed. By linearizing the circuit about the bias point, the sensitivities of each of the output variables to all the device values and model parameters will be calculated and output. This can easily generate huge amounts of output.

The *output_variable* has the same format and meaning as in the .PRINT statement for DC and transient analyses. However, in the case of the *output_variable* being a current, it is restricted to be the current through a voltage source.

Device sensitivities are provided for the following device types **only**: resistors, independent voltage and current sources, voltage- and current-controlled switches, diodes, and bipolar transistors.

.STEP   Parametric Analysis

General Forms

.STEP [LIN] *step_variable start end increment*

.STEP {DEC|OCT} *step_variable start end points*

.STEP *step_variable* LIST *value...*

Examples

```
.STEP VCE 0V 10V .5V
.STEP LIN I2 5mA -2mA 0.1mA
.STEP RES RMOD(R) 0.9 1.1 .001
.STEP DEC NPN QFAST(IS) 1E-18 1E-14 5
.STEP TEMP LIST 0 20 27 50 80 100
.STEP PARAM CenterFreq 9.5kHz 10.5kHz 50Hz
```

The .STEP statement causes a parametric sweep to be performed using *step_variable*, for all of the analyses of the circuit. The .STEP is at the same "level" as the .TEMP command: all of the other analyses (.AC, .DC, .TRAN, etc.) are done for each step. Each analysis generates an entire .PRINT table or .PLOT plot for each value of *step_variable*. Probe allows you to select any, or all, of the results of each analysis for merged display.

The .STEP statement is similar to the .DC statement, which raises the question of what happens if both .STEP and .DC commands try to set the same *variable*. This could also happen with Monte Carlo analysis. PSpice **does not allow** this: no two commands can try to set the same *variable*. This is flagged as an error during read-in and no analyses are done.

The value of *start* may be greater or less than *end*: that is, the stepping may proceed in either direction. The value of *increment* and *points* **must be positive**.

The stepping can be linear, logarithmic, or a list of values. The keyword LIN is optional for linear steps. The step types are

LIN   Linear step (sweep). The *step_variable* is stepped linearly from the value of *start* to *end*, and *increment* is the step size.

DEC   Log step (sweep), by decades. The *step_variable* is stepped logarithmically, by decades, from the value of *start* to *end*, and *points* is the number of steps per decade.

OCT   Log step (sweep), by octaves. The *step_variable* is stepped logarithmically, by octaves, from the value of *start* to *end*, and *points* is the number of steps per octave.

LIST   Use a list of values. In this case there are no start and/or end values. Instead, the numbers that follow the keyword LIST are the values to which *step_variable* will be set in sequence. Note: the values **must** be in either ascending or descending order.

The *step_variable* can be one of the following types:

Source: a name of an independent voltage, or current, source. During the sweep the source's voltage or current is set to the step value.

Model parameter: *model_type model_name (parameter_name)*
The parameter in the model is set to the step value. Note that model temperature parameters, such as TC1 and TC2 for the resistor, cannot be usefully stepped as temperature updates are processed before the .STEP stepping.

Temperature: use the keyword TEMP for *step_variable*. The temperature is set to the step value. For each value in the step, all the circuit components have their model parameters updated to that temperature.

Global parameter: use the keyword PARAM followed by the parameter name for *step_variable*. During the step, the global parameter's value is set to the step value and all expressions are reevaluated.

.STIMLIB   Stimulus Library File

General Form

```
.STIMLIB file_name
```

Examples

```
.STIMLIB mylib.stl
.STIMLIB normal
```

The .STIMLIB statement identifies stimulus library files created by StmEd and makes these available to the simulator. The file extension must be included.

.STIMULUS   Stimulus Definition

General Form

```
.STIMULUS name type parameters...
```

Examples

```
.STIMULUS  InPulse PULSE (-1mV 1mV 2nS 2nS 50nS 100nS)
.STIMULUS  DigPulse STIM (1,1)
+   0S    1
+   10NS  0
+   20NS  1
.STIMULUS  StdSin SIN (0 5 50K 0 0 0)
```

The .STIMULUS statement generally appear in library files created by StmEd. These statements define names that are referred to by analog source devices (I or V) or by the digital STIM device, for transient specifications **only**.

.SUBCKT   Subcircuit Definition

General Form

```
.SUBCKT name [[node ...]]
+ [[OPTIONAL: interface_node = default_node ...]]
+ [[PARAMS: name = value ...]]
+ [[TEXT: name = text_value ...]]
```

Examples

```
.SUBCKT  OPAMP  1  2  101  102  17
.SUBCKT  FILTER  INPUT, OUTPUT  PARAMS: CENTER=100kHz, WIDTH=10kHz
.SUBCKT  PLD IN1 IN2 IN3 OUT1
+  PARAMS: MNTYMXDLY=0 IO_LEVEL=0
+  TEXT:   JEDEC FILE="PROG.JED"
.SUBCKT  74LS00  A B Y
+  OPTIONAL: DPWR=$G_DPWR DGND=$G_DGND
+  PARAMS:   MNTYMXDLY=0 IO_LEVEL=0
```

The .SUBCKT statement begins the definition of a subcircuit. The definition is ended with a .ENDS statement. All the statements between .SUBCKT and .ENDS are included in the definition. Whenever the subcircuit is called, by an X statement, all the statements in the definition replace the calling statement.

The *name* is the subcircuit's name and is used by an X statement to reference the subcircuit. [[*node...*]] is an optional list of nodes (pins). There must be the same number of nodes in the subcircuit calling statements as in its definition. When the subcircuit is called, the actual nodes (the ones in the calling statement) replace the argument nodes (the ones in the defining statement). **Do not** use 0 ("zero") in this node list: that is reserved for global "ground" node.

The keyword OPTIONAL: allows you to specify optional nodes (pins) in the subcircuit definition. If an optional node is not specified in a subcircuit call (X statement), its *default_node* is used inside the subcircuit; otherwise, the *node* specified in the subcircuit call is used. This feature is particularly useful when specifying power supply nodes, because the same nodes are normally used in every device. This makes the subcircuits easier to use because the same nodes do not have to be specified in each subcircuit call.

The keyword PARAMS: allows values to be passed into subcircuits as arguments and used in expressions inside the subcircuit. See §10.7 (page 109) for more information on this capability.

The keyword TEXT: allows text values to be passed into subcircuits as arguments and used as expressions inside the subcircuit.

Subcircuit calls may be nested: an X statement may appear between a .SUBCKT and a .ENDS. However, subcircuit definitions **may not** be nested: this means that a .SUBCKT statement may not appear between a .SUBCKT and a .ENDS.

Subcircuit definitions should contain only device statements (statements without a leading ".") and possibly .FUNC, .MODEL, or .PARAM statements. Functions, models and parameters defined within a subcircuit definition are available **only** within the subcircuit definition in which they appear. Also, if a .FUNC, .MODEL, or .PARAM statement appears in the main circuit, that item is available in the main circuit **and** all subcircuits.

Node names (except for global nodes), device names, and model names are local to the subcircuit in which they are defined. It is OK to use a name in a subcircuit that has already been used in the main circuit. When the subcircuit is expanded all its names are prefixed by the subcircuit instance name: for example, Q13 becomes X3.Q13 and node 5 becomes X3.5 after expansion. After expansion all names are unique.

`.TEMP`   Temperature

General Form

> `.TEMP` *temperature...*

Examples

> ```
> .TEMP 125
> .TEMP 0 27 125
> ```

The `.TEMP` statement sets the temperature at which all analyses are done. The temperatures are in degrees Centigrade. If more than one temperature is given, then all analyses are done for each temperature.

It is assumed that the model parameters were measured or derived at the nominal temperature, `TNOM` (default 27°C), which may be set by the `.OPTIONS` statement.

`.TEXT`   Text Parameter Definition

General Form

> `.TEXT` *name* = { *"text"* | |*text_expression*| }...

Examples

> ```
> .TEXT MYFILE = "FILENAME.EXT"
> .TEXT FILE = "ROM.DAT", FILE2 = "ROM2.DAT"
> .TEXT PROGDAT = |"ROM"+TEXTINT(RUN_NO)+".DAT"| ;
> .TEXT DATA1 = "PLD.JED", PROGDAT = |"\PROG\DAT\"+FILENAME|
> ```

The keyword `.TEXT` is followed by a list of names with text values, which must be text constants (enclosed in "), or text expressions (enclosed in |). Text expressions may contain only text constants or previously defined text parameters. The `name` **cannot** be a `.PARAM` name, or any of the reserved `.PARAM` names.

A *text_expression* may contain:

> text constants (enclosed in ") and text parameters,

> the "+" operator (which concatenates two text values),

> the `TEXTINT(` { *value* | *expression* } `)` function, which returns a text string that is the nearest integer value to the value of its argument (interpreted as a floating-point value).

Once defined, a text parameter can be used as part of a text expression (as shown) or to specify: a JEDEC filename on a `PLD` device, an Intel "Hex" filename to program a `ROM` device or initialize a `RAM` device, a stimulus filename or signal name on an `FSTIM` device, or a text parameter to a subcircuit.

.TF   **Transfer Function**

General Form

> .TF *output_variable input_source*

Examples

```
.TF V(5) VIN
.TF I(VDRIV) ICNTRL
```

The .TF statement causes the small-signal transfer function to be calculated by linearizing the circuit around the bias point. The gain from *input_source* to *output_variable* will be output along with the input and output resistances. The output is done when these quantities are calculated and does not require .PRINT, .PLOT, or .PROBE statements.

The *output_variable* has the same format as in the .PRINT statement. However, if *output_variable* is a current, it must be the current through a voltage source.

.TRAN   **Transient Analysis**

General Form

> .TRAN[[/OP]] *print_step final_time* [[*results_delay* [[*step_ceiling*]]]] [[UIC]]

Examples

```
.TRAN     1ns 100ns
.TRAN/OP 1ns 100ns 20ns UIC
.TRAN     1ns 100ns 0ns .1ns
```

The .TRAN statement causes a transient analysis to be performed on the circuit. The transient analysis calculates the circuit's behavior over time, starting at TIME=0 and going to *final_time*.

The transient analysis uses an internal time step that is adjusted as the analysis proceeds. Over intervals where there is little activity, the internal time step is increased and during busy intervals it is decreased. The *print_step* is the time interval used for .PRINT and .PLOT results from the transient analysis. Since the analysis time-points are not the same as the print time-points, a 2nd-order polynomial interpolation is used to obtain the printed values.

The transient analysis always starts at TIME=0. However, it is possible to suppress output of the first portion of the analysis. The *results_delay* is the period for which no output is printed, plotted, or stored for Probe.

Sometimes you are concerned about the size of the internal time step. The default ceiling on the internal time step is *final_time*/50 (**it is not** *print_step*). The *step_ceiling* allows a ceiling smaller or larger than the print interval to be put on the internal time step.

Prior to doing the transient analysis, PSpice computes a bias point for the circuit separate from the regular bias point. This is done because an independent source can have a value at the start of transient analysis that is not the same as its DC value.

Normally only the node voltages are printed for the transient analysis bias point. However, the /OP suffix (on .TRAN) will cause the same detailed printing of the bias point that the .OP statement causes for the regular bias point.

If the keyword UIC (USE INITIAL CONDITIONS) is put at the end of the .TRAN statement, the calculation of the bias point is skipped. This option is used with the IC= specification for capacitors and inductors.

.WATCH   Watch Analysis Results

General Form

> .WATCH {AC|DC|TRAN} *output_variable* [[(*lower_limit*, *upper_limit*)]] ...

Examples

```
.WATCH DC V(3) (-1V,4V) V(2,3) V(R1)
.WATCH AC VM(2) VP(2) VMC(Q1)
.WATCH TRAN VBE(Q13) (0V,5V) ID(M2) I(VCC) (0,500mA)
.WATCH DC V([RESET]) (2.5V,10V)
```

The .WATCH statement allows results from AC, DC, and transient analyses to be output to the screen while the simulation is running. Up to three values can be seen on the display at one time. More than three *output_variables* can be specified, but they will not all be displayed. While only one analysis type may be specified per .WATCH statement, there may be a .WATCH statement for each analysis type in the circuit.

Following the analysis type is a list of the *output_variables* with optional value ranges. Up to eight output variables are allowed for a single .WATCH statement. The *output_variable* has the same format as in the .PRINT statement. The value range specifies the normal operating range of that particular output variable. If the range is exceeded during the simulation, the simulator will beep and pause. At this point, the simulation can be aborted or continued. If continued, the check for that output variable's boundary condition will be eliminated. Each output variable can have its own value range.

.WCASE   Sensitivity & Worst Case Analysis

General Form

.WCASE {AC|DC|TRAN} *output_value* *function* ⟦*option...*⟧

Examples

```
.WCASE TRAN V(5) YMAX
.WCASE DC IC(Q7) YMAX VARY DEV
.WCASE AC VP(13,5) YMAX DEVICES RQ OUTPUT ALL
```

The .WCASE statement causes a sensitivity and worst case analysis of the circuit. Multiple runs of the selected analysis (AC, DC, transient) are done while parameters are varied. Unlike .MC, .WCASE varies only one parameter per run. This allows PSpice to calculate the sensitivity of the output waveform to each parameter. Once all the sensitivities are calculated, one final run is done with all parameters varied so as to produce the user-specified, worst case waveform. The sensitivity and worst case runs are done with variations on model parameters as specified by the DEV and LOT tolerances on each .MODEL parameter (see the .MODEL statement for details on the DEV and LOT tolerances). The *output_value* has the same format as for specifying .PRINT output. Note: you can run either .MC or .WCASE **but not both** on the same circuit.

The other specifications on the .WCASE statement control the kind and amount of output generated by the Worst Case analysis. The *function* specifies the operation to be performed on the results from *output_value* to reduce this vector (a waveform of, for example, a voltage vs. time or a current vs. frequency, etc.) to a single value. This value is the basis for the comparisons between the nominal and subsequent runs. The *function* must be one of the following:

YMAX will find the **greatest difference** in each waveform from the nominal run.

MAX will find the **maximum value** of each waveform.

MIN will find the **minimum value** of each waveform.

RISE_EDGE(*value*) will find the **first occurrence** of the waveform crossing **above** the threshold *value*. The waveform must have one or more points at or below *value* followed by one above; the output value listed will be where the waveform rises above *value*.

FALL_EDGE(*value*) will find the **first occurrence** of the waveform crossing **below** the threshold *value*. The waveform must have one or more points at or above *value* followed by one below; the output value listed will be where the waveform falls below *value*.

The [[*option...*]] includes zero or more of the following:

LIST will print out, at the beginning of each run, the model parameter values used for each component during that run.

OUTPUT ALL requests output from the sensitivity runs, after the nominal (first) run. The content of the output from these runs is governed by the .PRINT, .PLOT, and .PROBE statements in the file. If OUTPUT ALL is omitted, then only the nominal and worst case (final) runs produce output.

RANGE(*low*, *high*) restricts the range over which *function* will be evaluated. A "*" can be used to indicate "for all values." For example:

YMAX RANGE(*,.5) will evaluate YMAX for the waveform for values of the sweep variable (time, frequency, etc.) of .5 or less.

MAX RANGE(-1,*) will find the maximum of the waveform for values of the sweep variable (time, frequency, etc.) of −1 or more.

If RANGE(*low*, *high*) is omitted, then *function* is evaluated over the whole sweep range. This is equivalent to RANGE(*, *).

{HI|LOW} specifies which direction the worst case run is to go relative to the nominal run. If *function* is YMAX or MAX the default is HI; otherwise the default is LOW.

VARY {DEV|LOT|BOTH} By default, any device that has a model parameter specifying either a DEV or a LOT tolerance will be included in the analysis. You may limit the analysis to only those devices that have DEV or LOT tolerances by specifying the appropriate option. The default is VARY BOTH.

DEVICES *device_type_list* By default, all devices are included in the sensitivity and worst case analyses. You may limit the devices considered by listing the device types to include in *device_type_list*. This list need not be in any particular order, but do **not** put any spaces, tabs, etc., in the list. For example, to only perform the analysis on resistors and MOSFETs, enter:

DEVICES RM

.WIDTH   **Width of Output**

General Form

```
.WIDTH OUT=value
```

Examples

```
.WIDTH OUT=80
.WIDTH OUT=132
```

Generally obsolete, the .WIDTH statement sets the width of the output file. The number of columns is set by *value*, which must be either 80 (the default) or 132.

This section summarizes the devices available in PSpice as of July, 1994. Each device is described by an example of its use in the circuit file, with some comments. Some devices have models with parameters, which are also listed. For the equations describing device currents, capacitances, temperature corrections, and noise currents, you may want to refer to the *Circuit Analysis Reference Manual* (available from MicroSim Corporation).

B

GaAsFET

General Form

B*name drain_node gate_node source_node model_name* [[*area*]]

Examples

```
BIN 100  1   0 GFAST
B13   22 14 23 GNOM 2.0
```

Model Form

.MODEL *model_name* GASFET([[*parameter = value*]]...)

Model Parameters (see the .MODEL statement)

parameter	Meaning	Units	Default
LEVEL	model index (1, 2, or 3)		1
VTO	pinch-off voltage	volt	−2.5
ALPHA	saturation voltage parameter	volt^{-1}	2
BETA	transconductance coefficient	amp/volt^2	0.1
B	doping tail extending parameter (LEVEL=2 only)	volt^{-1}	0.3
LAMBDA	channel-length modulation	volt^{-1}	0
GAMMA	static feedack parameter (LEVEL=3 only)		0
DELTA	output feedack parameter (LEVEL=3 only)	(amp·volt)$^{-1}$	0
Q	power-law parameter (LEVEL=3 only)		2
TAU	conduction current delay time	sec	0
RG	gate ohmic resistance	ohm	0
RD	drain ohmic resistance	ohm	0
RS	source ohmic resistance	ohm	0
IS	gate *p-n* saturation current	amp	1E−14

parameter	**Meaning**	**Units**	**Default**
N	gate *p-n* emission coefficient		1
M	gate *p-n* grading coefficient		0.5
VBI	gate *p-n* potential	volt	1
CGD	zero-bias gate-drain *p-n* capacitance	farad	0
CGS	zero-bias gate-source *p-n* capacitance	farad	0
CDS	drain-source capacitance	farad	0
FC	forward-bias depletion capacitance coefficient		0.5
VDELTA	capacitance transition voltage (LEVEL=2 or 3)	volt	0.2
VMAX	capacitance limiting voltage (LEVEL=2 or 3)	volt	0.5
EG	bandgap voltage (barrier height)	eV	1.11
XTI	IS temperature exponent		0
VTOTC	VTO temperature coefficient	volt/°C	0
BETATCE	BETA exponential temperature coefficient	%/°C	0
TRG1	RG temperature coefficient	°C^{-1}	0
TRD1	RD temperature coefficient	°C^{-1}	0
TRS1	RS temperature coefficient	°C^{-1}	0
KF	flicker noise coefficient		0
AF	flicker noise exponent		1
T_MEASURED	measured temperature (see .MODEL)	°C	
T_ABS	absolute temperature (see .MODEL)	°C	
T_REL_GLOBAL	relative to current temperature (see .MODEL)	°C	
T_REL_LOCAL	relative to AKO model temperature (see .MODEL)	°C	

The GaAsFET is modeled as an intrinsic FET with an ohmic resistance (RD/*area*) in series with the drain, another ohmic resistance (RS/*area*) in series with the source, and another ohmic resistance (RG) in series with the gate. [[*area*]] is the relative device area and defaults to 1.

Specific References

W. R. CURTICE, "A MESFET model for use in the design of GaAs integrated circuits," *IEEE Transactions on Microwave Theory and Techniques*, **MTT-28**, 448-456 (1980).

S. E. SUSSMAN-FORT, S. NARASIMHAN, and K. MAYARAM, "A complete GaAs MESFET computer model for SPICE," *IEEE Transactions on Microwave Theory and Techniques*, **MTT-32**, 471-473 (1984).

H. STATZ, P. NEWMAN, I. W. SMITH, R. A. PUCEL, and H. A. HAUS, "GaAs FET Device and Circuit Simulation in SPICE," *IEEE Transactions on Electron Devices*, **ED-34**, 160-169 (1987).

A. J. McCamant, G. D. McCormack, and D. H. Smith, "An Improved GaAs MESFET Model for SPICE," *IEEE Transactions on Microwave Theory and Techniques*, June 1990.

C   Capacitor

General Form

Cname +*node* −*node* ⟦*model_name*⟧ *value* ⟦IC=*initial_voltage*⟧

Examples

```
CLOAD   15   0 20pF
C2       1   2 .2E-12 IC=1.5V
CFDBCK   3  33 CMOD 10pF
```

Model Form

.MODEL *model_name* CAP(⟦*parameter* = *value*⟧...)

Model Parameters (see the .MODEL statement)

parameter	Meaning	Units	Default
C	capacitance multiplier		1
VC1	linear voltage coefficient	volt^{-1}	0
VC2	quadratic voltage coefficient	volt^{-2}	0
TC1	linear temperature coefficient	°C^{-1}	0
TC2	quadratic temperature coefficient	°C^{-2}	0
T_MEASURED	measured temperature (see .MODEL)	°C	
T_ABS	absolute temperature (see .MODEL)	°C	
T_REL_GLOBAL	relative to current temperature (see .MODEL)	°C	
T_REL_LOCAL	relative to AKO model temperature (see .MODEL)	°C	

The +*node* and −*node* define the polarity meant when the capacitor has a positive voltage across it. Positive current flows from the +*node* through the capacitor to the −*node*.

If ⟦*model_name*⟧ is left out, then *value* is the capacitance in farads. If ⟦*model_name*⟧ is specified, then the capacitance is given by the formula

$$value \cdot \text{C} \cdot \left(1 + \text{VC1} \cdot V_{CAP} + \text{VC2} \cdot V_{CAP}^2\right) \cdot \left(1 + \text{TC1} \cdot (T - T_{NOM}) + \text{TC2} \cdot (T - T_{NOM})^2\right)$$

where *value* is normally positive (though it can be negative, but **not** zero). T_{NOM} is the nominal temperature (set with TNOM in the .OPTIONS statement).

⟦IC=*initial_voltage*⟧ is the initial guess for the voltage across the capacitor during bias-point calculation. See §11.6 (page 126) for details on setting initial conditions.

Noise

The capacitor does not have a noise model.

D   **Diode**

General Form

> D*name* +*node* −*node* *model_name* [[*area*]]

Examples

```
DCLAMP 14  0 DMOD
D13    15 17 SWITCH 1.5
```

Model Form

> .MODEL *model_name* D([[*parameter* = *value*]]...)

Model Parameters (see the .MODEL statement)

parameter	Meaning	Units	Default
IS	saturation current	amp	1E−14
N	emission coefficient		1
ISR	recombination current parameter	amp	0
NR	emission coefficient for ISR		2
IKF	high-injection "knee" current	amp	*infinite*
BV	reverse breakdown "knee" voltage	volt	*infinite*
IBV	reverse breakdown "knee" current	amp	1E−10
NBV	reverse breakdown ideality factor		1
IBVL	low-level reverse breakdown "knee" current	amp	0
NBVL	low-level reverse breakdown ideality factor		1
RS	parasitic resistance	ohm	0
TT	transit time	second	0
CJO	zero-bias *p-n* capacitance	farad	0
VJ	*p-n* potential	volt	1
M	*p-n* grading coefficient		0.5
FC	forward-bias depletion capacitance coefficient		0.5
EG	bandgap voltage (barrier height)	eV	1.11
XTI	IS temperature exponent		3
TIKF	IKF temperature coefficient (linear)	°C^{-1}	0
TBV1	BV temperature coefficient (linear)	°C^{-1}	0
TBV2	BV temperature coefficient (quadratic)	°C^{-2}	0
TRS1	RS temperature coefficient (linear)	°C^{-1}	0
TRS2	RS temperature coefficient (quadratic)	°C^{-2}	0
KF	flicker noise coefficient		0
AF	flicker noise exponent		1
T_MEASURED	measured temperature (see .MODEL)	°C	

parameter	**Meaning**	**Units**	**Default**
T_ABS	absolute temperature (see .MODEL)	°C	
T_REL_GLOBAL	relative to current temperature (see .MODEL)	°C	
T_REL_LOCAL	relative to AKO model temperature (see .MODEL)	°C	

The diode is modeled as an ohmic resistance (RS/*area*) in series with an intrinsic diode. +*node* is the anode and −*node* is the cathode. Positive current is current flowing from the anode through the diode to the cathode. [[*area*]] scales the effect of IS, ISR, IKF, RS, CJO, IBV, and IBVL, and defaults to 1. BV, IBV, BVL, and IBVL are specified as positive values.

E　　　　　　　　**Voltage-Controlled Voltage Source**

General Forms

> E*name*　+*node*　−*node*　(+*input*, −*input*)　*gain*
>
> E*name*　+*node*　−*node*　POLY(*value*)　(+*input*, −*input*)... *coefficient*...
>
> E*name*　+*node*　−*node*　VALUE = {*expression*}
>
> E*name*　+*node*　−*node*　TABLE {*expression*} = (*input*, *output*)...
>
> E*name*　+*node*　−*node*　LAPLACE {*expression*} = {*transform_expression*}
>
> E*name*　+*node*　−*node*　FREQ {*expression*} = [[[MAG|DB]] [[DEG|RAD]]|R_I]]
> + (*freq*, *magnitude*, *phase*)... [[DELAY = *delay*]]
>
> E*name*　+*node*　−*node*　CHEBYSHEV {*expression*} =
> + {LP|HP|BP|BR}, *cutoff_freq*..., *attenuation*...

Examples

```
EBUFF     1    2  10,11   1.0
EAMP      13   0  POLY(1) 6,0  500
ENONLIN 100  101  POLY(2) 3,0  4,0  0.0 13.6 0.2 0.005
```

The first two general forms, linear and POLY, are part of SPICE. The VALUE, TABLE, LAPLACE, FREQ, and CHEBYSHEV general forms are part of PSpice's *Analog Behavioral Modeling* capability. Please refer to MicroSim's *Circuit Analysis Reference Manual* for details regarding these forms.

The +*node* and −*node* are the output nodes. Positive current flows from the +*node* through the source to the −*node*. The +*input* and −*input* nodes are in pairs and define a set of controlling voltages.

POLY(*value*) specifies the number of dimensions of the polynomial. The number of controlling node pairs must be equal to the number of dimensions. A particular node may appear more than once, and the output and controlling nodes need not be different.

The input to this device has infinite impedance; it draws no current. In addition, while the inputs sense the controlling node voltage(s), the device has no actual connection to those nodes. There is **no** DC path through the input nodes.

For the linear case, there are two controlling nodes followed by the *gain*. For the polynomial case, see §4.6 (page 31) for describing the controlling polynomial.

F   ## Current-Controlled Current Source

General Forms

> F*name* +*node* −*node* V*name* *gain*

> F*name* +*node* −*node* POLY(*value*) V*name*... *coefficient*...

Examples

```
FSENSE    1    2 VSENSE 10.0
FAMP      13   0 POLY(1) VIN  500
FNONLIN 100 101 POLY(2) VCNTRL1 VCNTRL2 0.0 13.6 0.2 0.005
```

The first general form and the first two examples apply to the linear case. The second general form and the last example are for the polynomial case.

The +*node* and −*node* are the output nodes. Positive current flows from the +*node* through the source to the −*node*. The current through the controlling voltage source determines the output current. The controlling source must be an independent voltage source (V device), although it need not have a zero DC value.

POLY(*value*) specifies the number of dimensions of the polynomial. The number of controlling voltage sources must be equal to the number of dimensions. A particular controlling source may appear more than once.

For the linear case, there is one controlling source followed by the *gain*. For the polynomial case, see §4.6 (page 31) for describing the controlling polynomial.

G   ## Voltage-Controlled Current Source

General Forms

> G*name* +*node* −*node* (+*input* , −*input*) *transconductance*

> G*name* +*node* −*node* POLY(*value*) (+*input* , −*input*)... *coefficient*...

> G*name* +*node* −*node* VALUE = {*expression*}

> G*name* +*node* −*node* TABLE {*expression*} = (*input* , *output*)...

> G*name* +*node* −*node* LAPLACE {*expression*} = {*transform_expression*}

> G*name* +*node* −*node* FREQ {*expression*} = [[MAG|DB]] [[DEG|RAD]] |R_I]]
> + (*freq* , *magnitude* , *phase*)... [DELAY = *delay*]]

> G*name* +*node* −*node* CHEBYSHEV {*expression*} =
> + {LP|HP|BP|BR}, *cutoff_freq*... , *attenuation*...

Examples

```
GBUFF      1    2  10,11  1.0
GAMP       13   0  POLY(1) 6,0   500
GNONLIN 100 101 POLY(2) 3,0  4,0  0.0 13.6 0.2 0.005
```

The first two general forms, linear and POLY, are part of SPICE. The VALUE, TABLE, LAPLACE, FREQ, and CHEBYSHEV general forms are part of PSpice's *Analog Behavioral Modeling* capability. Please refer to MicroSim's *Circuit Analysis Reference Manual* for details regarding these forms.

The +*node* and −*node* are the output nodes. Positive current flows from the +*node* through the source to the −*node*. The +*input* and −*input* nodes are in pairs and define a set of controlling voltages.

POLY(*value*) specifies the number of dimensions of the polynomial. The number of controlling node pairs must be equal to the number of dimensions. A particular node may appear more than once, and the output and controlling nodes need not be different.

The input to this device has infinite impedance; it draws no current. In addition, while the inputs sense the controlling node voltage(s), the device has no actual connection to those nodes. There is **no** DC path through the input nodes.

For the linear case, the two controlling nodes are followed by the "gain" or *transconductance*. For the polynomial case, see §4.6 (page 31) for describing the controlling polynomial.

H Current-Controlled Voltage Source

General Forms

H*name* +*node* −*node* V*name* *transresistance*

H*name* +*node* −*node* POLY(*value*) V*name*... *coefficient*...

Examples

```
HSENSE     1    2 VSENSE 10.0
HAMP       13   0 POLY(1) VIN   500
HNONLIN 100 101 POLY(2) VCNTRL1 VCNTRL2 0.0 13.6 0.2 0.005
```

The first general form and the first two examples apply to the linear case. The second general form and the last example are for the polynomial case.

The +*node* and −*node* are the output nodes. Positive current flows from the +*node* through the source to the −*node*. The current through the controlling voltage source determines the output current. The controlling source must be an independent voltage source (V device), although it need not have a zero DC value.

POLY(*value*) specifies the number of dimensions of the polynomial. The number of controlling voltage sources must be equal to the number of dimensions. A particular controlling source may appear more than once.

For the linear case, there is one controlling source followed by the "gain" or *transresistance*. For the polynomial case, see §4.6 (page 31) for describing the controlling polynomial.

I ⎯ Independent Current Source and Stimulus

General Form

> I*name* +*node* −*node* [[[DC]] *value*] [AC *magnitude* [*phase*]]
> + [STIMULUS = *name*]   [*transient_value*]

Examples

```
IBIAS   13  0   2.3mA
IAC      2  3   AC .001
IACPHS   2  3   AC .001 90
IPULSE   1  0   PULSE(-1mA 1mA 2ns 2ns 2ns 50ns 100ns)
I3      26 77   DC .002   AC 1   SIN(.002 .002 1.5MEG)
```

This element is a current source. Positive current flows from the +*node* through the source to the −*node*: in the first example, IBIAS drives node 13 to have a **negative** voltage. The default value is zero for the DC, AC, and transient values. None, any, or all of the DC, AC, and transient values may be specified. The AC [*phase*] is in degrees.

If present, the [*transient_value*] must be one of:

EXP *parameters*   for an exponential waveform (see page 117)

PULSE *parameters*   for a pulse waveform, which may repeat (see page 118)

PWL *parameters*   for a piecewise linear waveform (see page 119)

SFFM *parameters*   for a frequency-modulated waveform (see page 120)

SIN *parameters*   for a sinusoidal waveform (see page 121)

J ⎯ Junction FET

General Form

> J*name* *drain_node* *gate_node* *source_node* *model_name* [*area*]

Examples

```
JIN 100  1  0   JFAST
J13  22 14 23   JNOM  2.0
```

Model Forms

.MODEL *model_name* NJF ([*parameter* = *value*]...)

.MODEL *model_name* PJF ([*parameter* = *value*]...)

Model Parameters (see the .MODEL statement)

parameter	**Meaning**	**Units**	**Default**
VTO	threshold voltage	volt	−2
BETA	transconductance coefficient	amp/volt^2	1E−4
LAMBDA	channel-length modulation	volt^{-1}	0
IS	gate *p-n* saturation current	amp	1E−14
N	gate *p-n* emission coefficient		1
ISR	gate *p-n* recombination current parameter	amp	0
NR	emission coefficient for ISR		2
ALPHA	ionization coefficient	volt^{-1}	0
VK	ionization "knee" voltage	volt	0
RD	drain ohmic resistance	ohm	0
RS	source ohmic resistance	ohm	0
CGD	zero-bias gate-drain *p-n* capacitance	farad	0
CGS	zero-bias gate-source *p-n* capacitance	farad	0
M	gate *p-n* grading coefficient		0.5
PB	gate *p-n* potential	volt	1
FC	forward-bias depletion capacitance coefficient		0.5
VTOTC	VTO temperature coefficient	volt/°C	0
BETATCE	BETA exponential temperature coefficient	%/°C	0
XTI	IS temperature coefficient		3
KF	flicker noise coefficient		0
AF	flicker noise exponent		1
T_MEASURED	measured temperature (see .MODEL)	°C	
T_ABS	absolute temperature (see .MODEL)	°C	
T_REL_GLOBAL	relative to current temperature (see .MODEL)	°C	
T_REL_LOCAL	relative to AKO model temperature (see .MODEL)	°C	

The JFET is modeled as an intrinsic FET with an ohmic resistance (RD/*area*) in series with the drain, and with another ohmic resistance (RS/*area*) in series with the source. Positive current is current flowing into a terminal. ⟦*area*⟧ is the relative device area and defaults to 1.

Note: VTO < 0 means the device is a depletion-mode JFET (for both n-channel and p-channel) and VTO > 0 means the device is an enhancement-mode JFET. This conforms to U. C. Berkeley SPICE.

K   **Mutual Coupling**

General Forms

> K*name* L*name*... *coupling* [[*model_name* [[*size*]]]]

> K*name* T*name* T*name* Cm = *capacitive_coupling* Lm = *inductive_coupling*

Examples

```
KTUNED    L3OUT   L4IN  .8
KTRNSFRM  LPRIMARY  LSECNDRY .99
KXFRM     L1  L2  L3  L4 .98 KPOT_3C8
KLINES    T1  T2  Cm=.5pF  Lm=1mH
```

Model Form

> .MODEL *model_name* CORE([[*parameter* = *value*]]...)

Model Parameters (see the .MODEL statement)

parameter	Meaning	Units	Default
AREA	mean magnetic cross-section	cm^2	0.1
PATH	mean magnetic path length	cm	1
GAP	effective air-gap length	cm	0
PACK	pack (stacking) factor		1
MS	magnetization saturation	amp/meter	1E+6
A	thermal energy parameter	amp/meter	1E+3
C	domain flexing parameter		0.2
K	domain anisotropy parameter	amp/meter	500
ALPHA	interdomain coupling parameter		1E−3
GAMMA	domain damping parameter	sec^{-1}	*infinite*

Device Parameters (for transmission line coupling)

parameter	Meaning	Units	Default
Cm	capacitive coupling	farad/length	*none*
Lm	inductive coupling	henry/length	*none*

K*name* couples two, or more, inductors, or two transmission lines. Please refer to MicroSim's *Circuit Analysis Reference Manual* for details regarding this device.

L   **Inductor**

General Form

> *Lname* *+node* *−node* ⟦*model_name*⟧ *value* ⟦IC=*initial_current*⟧

Examples

```
LLOAD   15   0   20mH
L2       1   2   .2E-6
LCHOKE   3  42   LMOD .03
LSENSE   5  12   2UH  IC=2mA
```

Model Form

> .MODEL *model_name* IND(⟦*parameter = value*⟧...)

Model Parameters (see the .MODEL statement)

parameter	Meaning	Units	Default
L	inductance multiplier		1
IL1	linear current coefficient	amp^{-1}	0
IL2	quadratic current coefficient	amp^{-2}	0
TC1	linear temperature coefficient	°C^{-1}	0
TC2	quadratic temperature coefficient	°C^{-2}	0
T_MEASURED	measured temperature (see .MODEL)	°C	
T_ABS	absolute temperature (see .MODEL)	°C	
T_REL_GLOBAL	relative to current temperature (see .MODEL)	°C	
T_REL_LOCAL	relative to AKO model temperature (see .MODEL)	°C	

The *+node* and *−node* define the polarity meant when the inductor has a positive voltage across it. Positive current flows from the *+node* through the inductor to the *−node*.

If ⟦*model_name*⟧ is left out, then *value* is the inductance in henries. If ⟦*model_name*⟧ is specified, then the inductance is given by the formula

$$value \cdot L \cdot \left(1 + IL1 \cdot I_{IND} + IL2 \cdot I_{IND}^2\right) \cdot \left(1 + TC1 \cdot (T - T_{NOM}) + TC2 \cdot (T - T_{NOM})^2\right)$$

where *value* is normally positive (though it can be negative, but **not** zero). T_{NOM} is the nominal temperature (set with TNOM in the .OPTIONS statement).

⟦IC=*initial_current*⟧ is the initial guess for the current through the inductor during the bias-point calculation. See §11.6 (page 126) for more information on setting initial conditions.

Noise

The inductor does not have a noise model.

M   **MOSFET**

General Form

Mname *drain_node gate_node source_node bulk/substrate_node model_name*
+ [[L=*value*]] [[W=*value*]] [[AD=*value*]] [[AS=*value*]] [[PD=*value*]] [[PS=*value*]]
+ [[NRD=*value*]] [[NRS=*value*]] [[NRG=*value*]] [[NRB=*value*]] [[M=*value*]]

Examples

```
M1  14 2  13    0  PNOM   L=25u W=12u
M13 15 3  0     0  PSTRONG
M16 17 3  0     0  PSTRONG M=2
M28  0 2 100  100  NWEAK  L=33u W=12u
+   AD=288p AS=288p PD=60u PS=60u NRD=14 NRS=24 NRG=10
```

Model Forms

.MODEL *model_name* NMOS([[*parameter* = *value*]]...)

.MODEL *model_name* PMOS([[*parameter* = *value*]]...)

The MOSFET is modeled as an intrinsic MOSFET with ohmic resistances in series with the drain, source, gate, and bulk (substrate). There is also a shunt resistance (RDS) in parallel with the drain-source channel.

PSpice provides four MOSFET device models, which differ in the formulation of the I-V characteristic. The LEVEL parameter selects the different models (see §16.6, page 202).

L and W are the channel length and width, and are decreased to get the effective channel length and width. L and W can be specified in the device, model, or .OPTIONS statements. The value in the device statement supersedes the value in the model statement that supersedes the value in the .OPTIONS statement.

AD and AS are the drain and source diffusion areas. PD and PS are the drain and source diffusion perimeters. The drain-bulk and source-bulk saturation currents can be specified either by JS, which is multiplied by AD and AS, or by IS, which is an absolute value. The zero-bias depletion capacitances can be specified by CJ, which is multiplied by AD and AS, and by CJSW, which is multiplied by PD and PS. Or they can be set by CBD and CBS, which are absolute values.

NRD, NRS, NRG, and NRB are the relative resistivities of the drain, source, gate, and substrate in squares. These parasitic (ohmic) resistances can be specified either by RSH, which is multiplied by NRD, NRS, NRG, and NRB respectively or by RD, RS, RG, and RB, which are absolute values.

PD and PS default to 0, NRD and NRS default to 1, and NRG and NRB default to 0. Defaults for L, W, AD, and AS may be set in the .OPTIONS statement. If AD or AS defaults are not set, they also default to 0. If L or W defaults are not set, they default to 100μm.

M is a device "multiplier" (default = 1), which simulates the effect of multiple devices in parallel. The effective width, overlap and junction capacitances, and junction currents of the MOSFET are multiplied by M. The parasitic resistance values (RD, RS, etc.) are divided by M. Note the third example showing a device twice the size of the second example.

The DC characteristics of the first three model levels are defined by the parameters VTO, KP, LAMBDA, PHI, and GAMMA. These are computed by PSpice if process parameters (TOX, NSUB, etc.) are given, but the user-specified values always override. (Note: the default value for TOX is 0.1μm for model levels 2 and 3, but is unspecified for level 1; that "turns off" the use of process parameters.) VTO is positive (negative) for enhancement mode and negative (positive) for depletion mode of N-channel (P-channel) devices.

Model Parameters for Levels 1, 2, and 3 (see the .MODEL statement)

parameter	**Meaning**	**Units**	**Default**
LD	lateral diffusion (length)	meter	0
WD	lateral diffusion (width)	meter	0
VTO	zero-bias threshold voltage	volt	0
KP	transconductance coefficient	amp/volt^2	2E−5
LAMBDA	channel-length modulation (LEVEL=1 or 2)	volt^{-1}	0
PHI	surface potential	volt	0.6
GAMMA	bulk threshold parameter	volt$^{1/2}$	*calculated*
TOX	oxide thickness	meter	*see above*
TPG	gate material type: +1 = opposite of substrate, −1 = same as substrate, 0 = aluminum		+1
NSUB	substrate doping density	1/cm^3	none
NSS	surface state density	1/cm^2	none
NFS	fast surface state density	1/cm^2	0
XJ	metallurgical junction depth	meter	0
UO	(u-oh, **not** u-zero) surface mobility	cm^2/volt·sec	600
UCRIT	mobility degradation critical field (LEVEL=2)	volt/cm	1E+4
UEXP	mobility degradation exponent (LEVEL=2)		0
UTRA	mobility degradation transverse field coefficient		*not used*
VMAX	maximum drift velocity	meter/sec	0
NEFF	channel charge coefficient (LEVEL=2)		1
XQC	fraction of channel charge attributed to drain		1
DELTA	width effect on threshold		0
THETA	mobility modulation (LEVEL=3)	volt^{-1}	0
ETA	static feedback (LEVEL=3)		0
KAPPA	saturation field factor (LEVEL=3)		0.2

The LEVEL=4 (BSIM) model parameters are all values obtained from process characterization, and can be generated automatically. The last reference describes a means of generating a *process file*, which **must**, then, be converted into .MODEL statements for inclusion in a PSpice library or circuit file. (PSpice **does not** read process files.)

In the following list, parameters marked with a "∎" in the **L&W** column also have corresponding parameters with a length and width dependency. For example, VFB is a basic parameter with units of volts, and LVFB and WVFB also exist and have units of volt·μm. The equation

$$P_{effective} = P_{Basic} + \frac{P_{Length}}{L_{effective}} + \frac{P_{Width}}{W_{effective}} \tag{C-1}$$

is used to evaluate the parameter's effective value for an actual device, where $L_{effective} = L - DL$, and $W_{effective} = W - DW$.

Note that unlike the other models in PSpice, the BSIM model is designed for use with a process characterization system that provides all parameters: there are **no** defaults specified for the parameters, and leaving one out **may** cause problems.

Model Parameters for Level 4 (see the .MODEL statement)

parameter	Meaning	Units	L&W
DL	channel shortening	μm	
DW	channel narrowing	μm	
TOX	gate-oxide thickness	μm	
VFB	flat-band voltage	volt	∎
PHI	surface inversion potential	volt	∎
K1	body effect coefficient	volt^{½}	∎
K2	drain/source depletion charge sharing coefficient		∎
ETA	zero-bias drain-induced barrier lowering coefficient		∎
X2E	sensitivity of drain-induced barrier lowering effect to substrate bias	volt^{-1}	∎
X3E	sensitivity of drain-induced barrier lowering effect to drain bias at $V_{DS} = V_{DD}$	volt^{-1}	∎
MUZ	zero-bias mobility	cm^2/volt·sec	
X2MZ	sensitivity of mobility to substrate bias at $V_{DS} = 0$	cm^2/volt^2·sec	∎
U0	zero-bias transverse-field mobility degradation	volt^{-1}	∎
X2U0	sensitivity of transverse-field mobility degradation effect to substrate bias	volt^{-2}	∎
U1	zero-bias velocity saturation	μm/volt	∎
X2U1	sensitivity of velocity saturation effect to substrate bias	μm/volt^2	∎
X3U1	sensitivity of velocity saturation effect on drain bias at $V_{DS} = V_{DD}$	μm/volt^2	∎
MUS	mobility at zero substrate bias and at $V_{DS} = V_{DD}$	cm^2/volt^2·sec	∎
X2MS	sensitivity of mobility to substrate bias at $V_{DS} = 0$	cm^2/volt^2·sec	∎
X3MS	sensitivity of mobility to drain bias at $V_{DS} = V_{DD}$	cm^2/volt^2·sec	∎

parameter	Meaning	Units	L&W
N0	zero-bias subthreshold slope coefficient		■
NB	sensitivity of subthreshold slope to substrate bias		■
ND	sensitivity of subthreshold slope to drain bias		■
TEMP	temperature at which parameters were measured	°C	
VDD	measurement bias range		
XPART	gate-oxide capacitance charge model flag: XPART=0 selects 40:60 drain:source charge partition, XPART=1 selects 0:100 drain:source charge partition.		
WDF	drain, source junction default width	meter	
DELL	drain, source junction length reduction	meter	

The LEVEL=5 (BSIM3) model is a physical model with extensive built-in dependencies of important dimensional and processing parameters. It includes the major effects that are important to modeling deep-submicron MOSFETs, such as threshold voltage reduction, non-uniform doping, mobility reduction due to the vertical field, bulk charge effect, carrier velocity saturation, drain-induced barrier lowering, channel length modulation, hot-carrier induced output-resistance reduction, etc. Please refer to MicroSim's *Circuit Analysis Reference Manual* for details regarding this model.

The following list describes the parameters common to all model levels, which are primarily parasitic element values such as series resistance, overlap and junction capacitances, and so on.

Model Parameters for All Levels (see the .MODEL statement)

parameter	Meaning	Units	Default
LEVEL	model index		1
L	channel length	meter	DEFL
W	channel width	meter	DEFW
RD	drain ohmic resistance	ohm	0
RS	source ohmic resistance	ohm	0
RG	gate ohmic resistance	ohm	0
RB	bulk ohmic resistance	ohm	0
RDS	drain-source shunt resistance	ohm	*infinite*
RSH	drain, source diffusion sheet resistance	ohm/square	0
IS	bulk *p-n* saturation current	amp	1E−14
JS	bulk *p-n* saturation current per unit area	amp/meter^2	0
JSSW	bulk *p-n* saturation sidewall current per unit length	amp/meter	0
N	bulk *p-n* emission coefficient		1
PB	bulk *p-n* bottom potential	volt	0.8
PBSW	bulk *p-n* sidewall potential	volt	PB
CBD	zero-bias bulk-drain *p-n* capacitance	farad	0

parameter	Meaning	Units	Default
CBS	zero-bias bulk-source *p-n* capacitance	farad	0
CJ	bulk *p-n* zero-bias bottom capacitance per unit area	farad/meter^2	0
CJSW	bulk *p-n* zero-bias sidewall capacitance per unit length	farad/meter	0
MJ	bulk *p-n* bottom grading coefficient		0.5
MJSW	bulk *p-n* sidewall grading coefficient		0.33
FC	bulk *p-n* forward-bias capacitance coefficient		0.5
TT	bulk *p-n* transit time	sec	0
CGSO	gate-source overlap capacitance per unit channel width	farad/meter	0
CGDO	gate-drain overlap capacitance per unit channel width	farad/meter	0
CGBO	gate-bulk overlap capacitance per unit channel length	farad/meter	0
KF	flicker noise coefficient		0
AF	flicker noise exponent		1
T_MEASURED	measured temperature (see .MODEL)	°C	
T_ABS	absolute temperature (see .MODEL)	°C	
T_REL_GLOBAL	relative to current temperature (see .MODEL)	°C	
T_REL_LOCAL	relative to AKO model temperature (see .MODEL)	°C	

Specific References

H. SHICHMAN and D. A. HODGES, "Modeling and simulation of insulated-gate field-effect transistor switching circuits," *IEEE Journal of Solid-State Circuits*, **SC-3**, 285, September 1968.

B. J. SHEU, D. L. SCHARFETTER, P.-K. KO, and M.-C. JENG, "BSIM: Berkeley Short-Channel IGFET Model for MOS Transistors," *IEEE Journal of Solid-State Circuits*, **SC-22**, 558-566, August 1987.

P. ANTOGNETTI and G. MASSOBRIO, *Semiconductor Device Modeling with SPICE*, McGraw-Hill, 1993.

A. VLADIMIRESCU, and S. LIU, *The Simulation of MOS Integrated Circuits Using SPICE2*, Memorandum No. M80/7, February 1980.

J. R. PIERRET, *A MOS Parameter Extraction Program for the BSIM Model*, Memorandum No. M84/99 and M84/100, November 1984.

The last two references are available for US$10 (each) by sending a check payable to *The Regents of the University of California* to this address:

Cindy Manly-Fields
EECS/ERL Industrial Support Office
497 Cory Hall
University of California
Berkeley, CA 94720

N   Digital Input

General Form

> N*name interface_mode low_level_node high_level_node model_name*
> + { DGTLNET= *digital_net IO_model_name* | [[SIGNAME= *signal_name*]] }
> + [[IS= *initial_state*]]

Examples

```
N1   ANALOG  D_GND  D_PWR  DIN74  DGTLNET=DIGITAL_NODE   IO_STD
NRESET  7 15  16  FROM_TTL
N12    18  0 100  FROM_CMOS  SIGNAME=VCO_GATE   IS=0
```

Model Form

> .MODEL *model_name* DINPUT([[*parameter* = *value*]]...)

Please refer to MicroSim's *Circuit Analysis Reference Manual* for details regarding this device.

O   Digital Output

General Form

> O*name interface_mode reference_node model_name*
> + { DGTLNET= *digital_net IO_model_name* | [[SIGNAME= *signal_name*]] }

Examples

```
O12 ANALOG_NODE DIGITAL_GND DO74 DGTLNET=DIGITAL_NODE IO_STD
OVCO 17   0 TO_TTL
O5    22 100 TO_CMOS SIGNAME=VCO_OUT
```

Model Form

> .MODEL *model_name* DOUTPUT([[*parameter* = *value*]]...)

Please refer to MicroSim's *Circuit Analysis Reference Manual* for details regarding this device.

Q   **Bipolar Transistor**

General Form

> *Qname collector_node base_node emitter_node [[substrate_node]]*
> + *model_name [[area]]*

Examples

```
Q1   14 2 13 PNPNOM
Q13 15 3  0 1 NPNSTRONG 1.5
Q7   VC 5 12 [SUB] LATPNP
```

Model Forms

.MODEL *model_name* NPN([[*parameter = value*]]...)

.MODEL *model_name* PNP([[*parameter = value*]]...)

.MODEL *model_name* LPNP([[*parameter = value*]]...)

Model Parameters (see the .MODEL statement)

parameter	Meaning	Units	Default
IS	transport saturation current	amp	1E−16
BF	ideal maximum forward beta		100
NF	forward current emission coefficient		1
VAF *or* VA	forward Early voltage	volt	*infinite*
IKF *or* IK	forward-beta high-current roll-off "knee" current	amp	*infinite*
ISE *or* C2	base-emitter leakage saturation current	amp	0
NE	base-emitter leakage emission coefficient		1.5
BR	ideal maximum reverse beta		1
NR	reverse current emission coefficient		1
VAR *or* VB	reverse Early voltage	volt	*infinite*
IKR	corner for reverse-beta high-current roll-off	amp	*infinite*
ISC *or* C4	base-collector leakage saturation current	amp	0
NC	base-collector leakage emission coefficient		2
NK	high-current roll-off coefficient		0.5
ISS	substrate *p-n* saturation current	amp	0
NS	substrate *p-n* emission coefficient		1
RE	emitter ohmic resistance	ohm	0
RB	zero-bias (maximum) base resistance	ohm	0
RBM	minimum base resistance	ohm	RB
IRB	current at which R_B falls halfway to RBM	amp	*infinite*
RC	collector ohmic resistance	ohm	0
CJE	base-emitter zero-bias *p-n* capacitance	farad	0

parameter	Meaning	Units	Default
VJE *or* PE	base-emitter built-in potential	volt	0.75
MJE *or* ME	base-emitter *p-n* grading factor		0.33
CJC	base-collector zero-bias *p-n* capacitance	farad	0
VJC *or* PC	base-collector built-in potential	volt	0.75
MJC *or* MC	base-collector *p-n* grading factor		0.33
XCJC	fraction of C_{BC} connected internal to R_B		1
CJS *or* CCS	substrate zero-bias *p-n* capacitance	farad	0
VJS *or* PS	substrate *p-n* built-in potential	volt	0.75
MJS *or* MS	substrate *p-n* grading factor		0
FC	forward-bias depletion capacitor coefficient		0.5
TF	ideal forward transit time	sec	0
XTF	transit time bias dependence coefficient		0
VTF	transit time dependency on V_{BC}	volt	*infinite*
ITF	transit time dependency on I_C	amp	0
PTF	excess phase at $1/(2\pi \cdot TF)$Hz	degree	0
TR	ideal reverse transit time	sec	0
QCO	epitaxial region charge factor	coulomb	0
RCO	epitaxial region resistance	ohm	0
VO	carrier mobility "knee" voltage	volt	10
GAMMA	epitaxial region doping factor		1E−11
EG	bandgap voltage (barrier height)	eV	1.11
XTB	forward and reverse beta temperature coefficient		0
XTI *or* PT	IS temperature effect exponent		3
TRE1	RE temperature coefficient (linear)	°C^{-1}	0
TRE2	RE temperature coefficient (quadratic)	°C^{-2}	0
TRB1	RB temperature coefficient (linear)	°C^{-1}	0
TRB2	RB temperature coefficient (quadratic)	°C^{-2}	0
TRM1	RBM temperature coefficient (linear)	°C^{-1}	0
TRM2	RBM temperature coefficient (quadratic)	°C^{-2}	0
TRC1	RC temperature coefficient (linear)	°C^{-1}	0
TRC2	RC temperature coefficient (quadratic)	°C^{-2}	0
KF	flicker noise coefficient		0
AF	flicker noise exponent		1
T_MEASURED	measured temperature (see .MODEL)	°C	
T_ABS	absolute temperature (see .MODEL)	°C	
T_REL_GLOBAL	relative to current temperature (see .MODEL)	°C	
T_REL_LOCAL	relative to AKO model temperature (see .MODEL)	°C	

The bipolar transistor is modeled as an intrinsic transistor with ohmic resistances in series with the collector (RC/*area*), the base (value varies with current), and the emitter (RE/*area*). Positive current is current flowing into a terminal. [[*area*]] is the relative device area and defaults to 1. For those model parameters that have alternate names, such as VAF and VA, either name may be used.

The substrate node is optional, and if not specified it defaults to ground. Because PSpice allows alphanumeric names for nodes, and because there is no easy way to distinguish these from the model names, it is necessary to enclose the name (not a number) used for the substrate node with square brackets; otherwise it will be interpreted as a model name. See the third example.

For model types NPN and PNP, the isolation junction capacitance is connected between the intrinsic-collector and substrate nodes. This is the same as in SPICE and works well for vertical IC transistor structures. For lateral IC transistor structures there is a third model, LPNP, where the isolation junction capacitance is connected between the intrinsic-base and substrate nodes.

The parameters ISE (or C2) and ISC (or C4) may be set to be greater than 1. In this case, they are interpreted as multipliers of IS instead of as absolute currents: if ISE > 1, then it is replaced by ISE·IS. Likewise for ISC.

If the model parameter RCO is specified, then quasi-saturation effects are included.

Specific References

IAN GETREU, *Modeling the Bipolar Transistor*, Tektronix, Inc. part no. 062-2841-00.

G. M. KULL, L. W. NAGEL, S. W. LEE, P. LLOYD, E. J. PRENDERGAST, and H. K. DIRKS, "A Unified Circuit Model for Bipolar Transistors Including Quasi-Saturation Effects," *IEEE Transactions on Electron Devices*, **ED-32**, 1103-1113 (1985).

R # Resistor

General Form

Rname +node −node [[model_name]] value [[TC=TC1 [[,TC2]]]]

Examples

```
RLOAD 15 0 2K
R2    1 2 2.4E4 TC=.015,-.003
RFDBCK 3 33 RMOD 10K
```

Model Form

.MODEL model_name RES([[parameter = value]]...)

Model Parameters (see the .MODEL statement)

parameter	Meaning	Units	Default
R	resistance multiplier		1
TC1	linear temperature coefficient	°C^{-1}	0
TC2	quadratic temperature coefficient	°C^{-2}	0
TCE	exponential temperature coefficient	%/°C	0
T_MEASURED	measured temperature (see .MODEL)	°C	
T_ABS	absolute temperature (see .MODEL)	°C	
T_REL_GLOBAL	relative to current temperature (see .MODEL)	°C	
T_REL_LOCAL	relative to AKO model temperature (see .MODEL)	°C	

The +node and −node define the polarity meant when the resistor has a positive voltage across it. Positive current flows from the +node through the resistor to the −node.

Temperature coefficients for the resistor can be specified in-line, as in the second example. If the resistor **has** a model specified, then the coefficients from the model are used for the temperature updates; otherwise the in-line values are used. In both cases the temperature coefficients default to zero. Expressions may **not** be used for the in-line coefficients.

If [[model_name]] is left out, then value is the resistance in ohms. If [[model_name]] is specified and TCE (in the model) **is not** specified, then the resistance is given by the formula

$$value \cdot R \cdot \left(1 + TC1 \cdot (T - T_{NOM}) + TC2 \cdot (T - T_{NOM})^2\right)$$

If [[model_name]] is specified and TCE (in the model) **is** specified, then the resistance is given by the formula

$$value \cdot R \cdot 1.01^{TCE \cdot (T - T_{NOM})}$$

In both cases, *value* is normally positive (though it can be negative, but **not** zero) and T_{NOM} is the nominal temperature (set with TNOM in the .OPTIONS statement).

S   **Voltage-Controlled Switch**

General Form

Sname +node −node +input −input model_name

Examples

```
S12     13 17  2 0 SMOD
SRESET   5  0 15 3 RELAY
```

Model Form

.MODEL model_name VSWITCH([[parameter = value]]...)

Model Parameters (see the .MODEL statement)

parameter	Meaning	Units	Default
RON	"on" resistance	ohm	1
ROFF	"off" resistance	ohm	1E+6
VON	control voltage for "on" state	volt	1
VOFF	control voltage for "off" state	volt	0

The voltage-controlled switch is a special kind of voltage-controlled resistor. The resistance between +*node* and −*node* depends on the voltage between the +*input* and −*input*. The resistance varies continuously between RON and ROFF.

RON and ROFF must be greater than zero and less than 1/GMIN. A resistance of 1/GMIN is connected between the controlling nodes to keep them from floating. (Set with GMIN in the .OPTIONS statement.)

This model for a switch minimizes numerical problems. However, there are a few things to keep in mind:

- With double precision numbers, PSpice can only handle a dynamic range of about 12 decades. It is recommended that the ratio of ROFF to RON be less than 1E+12.

- Similarly, it is recommended that the transition region not be too narrow. Remember that in the transition region the switch has gain. The narrower the region, the higher the gain and the greater the potential for numerical problems. The minimum value allowed for VON−VOFF is:

    RELTOL × *max*(VON,VOFF) + VNTOL

- Although very little computer time is required to evaluate switches, during transient analysis PSpice must step through the transition region with a step size small enough to produce an accurate waveform. If there are many transitions you may have long run times from simulating the rest of the circuit through these transitions.

T   **Transmission Line**

General Forms: ideal

> T*name* +A_port −A_port +B_port −B_port Z0=*impedance*
> + { TD=*delay* | F=*frequency* [[NL=*wavelength*]] }
> + IC=*A_voltage*, *A_current*, *B_voltage*, *B_current*

> T*name* +A_port −A_port +B_port −B_port *model*
> + IC=*A_voltage*, *A_current*, *B_voltage*, *B_current*

General Forms: lossy

> T*name* +A_port −A_port +B_port −B_port LEN=*electrical_length*
> + R=*resistance_per_length* L=*inductance_per_length*
> + G=*conductance_per_length* C=*capacitance_per_length*

> T*name* +A_port −A_port +B_port −B_port *model* [[*electrical_length*]]

Examples

```
T1   1 2   3 4   Z0=220 TD=115ns
T2   1 2   3 4   Z0=220 F=2.25MHz
T3   1 2   3 4   Z0=220 F=4.5MHz NL=0.5
T4   1 2   3 4   LEN=1 R=.31 L=.38uH G=6.3uM C=67pF
```

Model Form

> .MODEL *model_name* TRN([[*parameter = value*]]...)

Model Parameters (see the .MODEL statement)

parameter	Meaning	Units	Default
Z0	**ideal**: characteristic impedance	ohm	
TD	**ideal**: transmission delay	sec	
F	**ideal**: frequency for NL	Hertz	
NL	**ideal**: relative wavelength		0.25
LEN	**lossy**: electrical length	*user*	
R	**lossy**: resistance per LEN's units	ohm/unit	
L	**lossy**: inductance per LEN's units	henry/unit	
G	**lossy**: conductance per LEN's units	mho/unit	
C	**lossy**: capacitance per LEN's units	farad/unit	

The transmission line device is a bidirectional, ideal delay line. It has two ports, that we will refer to as A and B. The "+" and "−" port nodes define the polarity of a positive voltage at that port.

Z0 is the characteristic impedance. The transmission line's length can be specified either by TD, a delay in seconds, or by F and NL, a frequency and a relative wavelength at the specified frequency. NL defaults to 1/4 (F is then the quarter-wave frequency). Note that the first set of examples all specify the same transmission line. The `IC=` sets the initial guess for voltage across and current into the ports.

During transient analysis, the internal time step is limited to be no more than one-half the smallest transmission delay, so short transmission lines will cause long run times.

Note: both `Z0` ("zee-zero") and `ZO` ("zee-oh") are accepted by PSpice.

U **Digital Device**

General Forms

Uname *primitive_type* ([[*parameter*]]...) *node...*
+ *timing_model_name IO_model_name*

Uname STIM(*width* , *format*) *node...*
+ *IO_model_name* [[TIMESTEP=*stepsize*]] *waveform_description*

Examples

```
U1 NAND(2) 1 2 10 D0_GATE IO_DFT
U2 JKFF(1) 3 5 200 3 3 10 2 D_293ASTD IO_STD
U3 STIM(1, 1) 110 STMIOMDL TIMESTEP = 10ns
+   0ns, 1
+   40ns, 0
```

IO Model Form

.MODEL *model_name* UIO([[*parameter = value*]]...)

Timing Model Forms

Standard Gates
.MODEL *model_name* UGATE([[*parameter = value*]]...)

Tri-State Gates
.MODEL *model_name* UTGATE([[*parameter = value*]]...)

Bidirectional Transfer Gates
.MODEL *model_name* UBTG([[*parameter = value*]]...)

Edge-Triggered Flip-Flops
.MODEL *model_name* UEFF([[*parameter = value*]]...)

Gated Flip-Flops
.MODEL *model_name* UGFF([[*parameter = value*]]...)

Digital Delay Lines
.MODEL *model_name* UDLY([[*parameter = value*]]...)

Programmable Logic Arrays
.MODEL *model_name* UPLD([[*parameter = value*]]...)

Read-Only Memories
    .MODEL *model_name* UROM([[*parameter* = *value*]]...)

Random Access (read/write) Memories
    .MODEL *model_name* URAM([[*parameter* = *value*]]...)

Multi-bit Analog-to-Digital Converters
    .MODEL *model_name* UADC([[*parameter* = *value*]]...)

Multi-bit Digital-to-Analog Converters
    .MODEL *model_name* UDAC([[*parameter* = *value*]]...)

Please refer to MicroSim's *Circuit Analysis Reference Manual* for details regarding this device.

V   Independent Voltage Source and Stimulus

General Form

> V*name* +*node* −*node* [[[DC]] *value*] [AC *magnitude* [*phase*]]
> + [STIMULUS = *name*] [*transient_value*]

Examples

```
VBIAS   13   0   2.3mV
VAC      2   3   AC .001
VACPHS   2   3   AC .001 90
VPULSE   1   0   PULSE(-1mV 1mV 2ns 2ns 2ns 50ns 100ns)
V3      26  77   DC .002   AC 1   SIN(.002 .002 1.5MEG)
```

This element is a voltage source. Positive current (as recorded in the output file, or for Probe, or as used as an input to a current-controlled device) flows into the +*node* through the source and out of the −*node*. The default value is zero for the DC, AC, and transient values. None, any, or all of the DC, AC, and transient values may be specified. The AC [*phase*] is in degrees.

If present, the [*transient_value*] must be one of:

EXP *parameters*	for an exponential waveform (see page 117)
PULSE *parameters*	for a pulse waveform, which may repeat (see page 118)
PWL *parameters*	for a piecewise linear waveform (see page 119)
SFFM *parameters*	for a frequency-modulated waveform (see page 120)
SIN *parameters*	for a sinusoidal waveform (see page 121)

W   Current-Controlled Switch

General Form

   Wname +node −node Vname model_name

Examples

      W12   13 17 Vctrl WMOD
      WRESET   5 0 Vreset RELAY

Model Form

   .MODEL *model_name* ISWITCH([[*parameter* = *value*]]...)

Model Parameters (see the .MODEL statement)

parameter	Meaning	Units	Default
RON	"on" resistance	ohm	1
ROFF	"off" resistance	ohm	1E+6
ION	control current for "on" state	amp	1
IOFF	control current for "off" state	amp	0

The current-controlled switch is a special kind of voltage-controlled resistor. The resistance between +*node* and −*node* depends on the current through the voltage source *Vname*. The resistance varies continuously between RON and ROFF.

RON and ROFF must be greater than zero and less than 1/GMIN. A resistance of 1/GMIN is connected between the controlling nodes to keep them from floating. (Set with GMIN in the .OPTIONS statement.)

This model for a switch minimizes numerical problems. However, there are a few things to keep in mind:

- With double precision numbers, PSpice can only handle a dynamic range of about 12 decades. It is recommended that the ratio of ROFF to RON be less than 1E+12.

- Similarly, it is recommended that the transition region not be too narrow. Remember that in the transition region the switch has gain. The narrower the region, the higher the gain and the greater the potential for numerical problems. The minimum value allowed for ION−IOFF is:

   RELTOL × *max*(ION,IOFF) + ABSTOL

- Although very little computer time is required to evaluate switches, during transient analysis PSpice must step through the transition region with a step size small enough to produce an accurate waveform. If there are many transitions you may have long run times from simulating the rest of the circuit through these transitions.

X   Subcircuit Call

General Form

> X*name* [[*node* ...]] *subcircuit_name*
> + [[PARAMS: *name* = *value* ...]]
> + [[TEXT: *name* = *text_value* ...]]

Examples

```
X12   100  101  200  201  DIFFAMP
XBUFF  13  15  UNITAMP
```

The *subcircuit_name* is the name of the subcircuit's definition (see the .SUBCKT statement). There must be the same number of nodes in the instance (or call) as in the subcircuit's definition. This statement causes the referenced subcircuit to be inserted into the circuit with the given nodes replacing the argument nodes in the definition. It allows you to define a block of circuitry once and then use that block in several places.

The keyword PARAMS: allows values to be passed into subcircuits as arguments and used in expressions inside the subcircuit.

The keyword TEXT: allows text values to be passed into subcircuits and used in text expressions inside the subcircuit.

Subcircuit calls may be nested. That is, you may have a call to subcircuit A, whose definition contains a call to subcircuit B. The nesting may be to any level, but **must not** be circular: for example, if subcircuit A's definition contains a call to subcircuit B, then subcircuit B's definition must not contain a call to subcircuit A.

USING THE CONTROL SHELL, PROBE, AND THE STIMULUS EDITOR

This section provides some references, and a brief discussion of running the Control Shell, Probe, and the Stimulus Editor (StmEd). These directions are meant as an introduction only, and you should always refer to MicroSim Corporation's product documentation for the breadth and detail of using these programs. These products are designed to be easy to use, which implies that they are easy to learn; for the most part, you should be able to "follow your nose" and discover the appropriate sequence of menus and commands to accomplish most operations.

D.1 USING THE CONTROL SHELL

Available **only** for the IBM-PC, the Control Shell (or "shell") acts as a "hub" for running your simulations. It is **not** necessary that you use the shell, but you may find it convenient. The shell program is an executable file, PSHELL.EXE, but this file is **not** meant to be run directly. Instead, there is a small, supervisory program in the

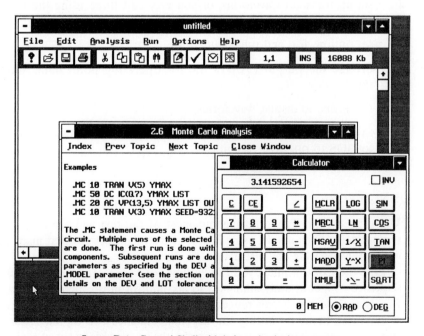

FIGURE D-1 Control Shell with help and calculator popup.

executable file PS.EXE that runs all of the other programs with the correct "switch" settings and filename arguments. So, it is intended that you use the following command (at the DOS prompt):

    PS [[–m]] [[*filename*]]

where [[*filename*]] is the name of the circuit (.CIR) file you are simulating, but this is an option as you may select this from within the shell. The [[–m]] option is for computers using monochrome video.

The menus are easy to select and activate; you can use the keyboard and/or a Microsoft-compatible mouse. Note the following points:

- Only the highlighted menu items are available.

- Menus and their entries may be selected using the cursor keys, the mouse, or by hitting the key for the letter of the item you are selecting.

- The reverse-video entry will be activated by hitting Enter or clicking the mouse.

- The escape key (Esc) is an "undo" for most menu selections.

Context-sensitive help is available for most subjects. Many interactive features are available within the Control Shell. See the manuals from MicroSim for these details.
Using the shell, you can:

- create and modify your circuit file, which may include running StmEd to set-up transient waveforms, or you may edit these using the fill-in form screens in the shell itself,

- set options for, and run, simulations,

- browse the output (.OUT) file for operating point data, accounting information, etc.,

- run Probe to display waveforms,

- save your circuit file, start a new circuit, visit previous work, or shell-out to DOS to run other programs.

D.2 USING PROBE

We've been using Probe throughout most of the text and are only now reviewing some details. Fortunately, using Probe is so straightforward that many people use it without ever seeing any instructions.

Like the Control Shell, you can "follow your nose" and discover the appropriate sequence of menus and commands to accomplish most operations. For example, if you want to set the X-axis range, select X_axis, and then Set_range. You may select these items by hitting the key that matches the capital letter in the item (usually the first letter), or you may move the reverse-video select region to the item with the cursor keys and then hit Enter, or you may click on the item with the mouse cursor. If you use a mouse, it will need to be a Microsoft-compatible mouse.

Probe will work with many displays and hardcopy devices. The configuration for these is stored in the file PSPICE.DEV, where it can be accessed by Probe, StmEd, and Parts (another MicroSim product). The shell lets you edit PSPICE.DEV directly, assuming that you know the ciphers for the displays and hardcopy devices supported by Probe. An easy way to avoid problems is to use a program called "SetupDev," which sets-up these devices.

The SetupDev program is an executable file, SETUPDEV.EXE, that runs your computer like a "glass teletype" or "dumb terminal" (whichever phrase you prefer). You should select from the options it presents, review your selections, and be sure to save the selections before quitting the program.

The student version of Probe is limited to using the results from the student version of PSpice. Also, the DOS student version shares the DOS professional version limit of about 8,000 data points per waveform on the screen. You can still see all sections of a longer waveform but only 8,000 points at a time. If this limit occurs, Probe will prompt you to select the section of the waveform you want to inspect.

D.3 USING THE STIMULUS EDITOR

The Stimulus Editor is like Probe, except that the waveforms it displays are the stimuli for your circuit. Stimulus Editor, or StmEd (pronounced "stim-ed") as its menu reads in the Control Shell, is contained in the executable file STMED.EXE as a separate program. StmEd may be started from the shell or, like Probe, it may be run directly:

> STMED *circuit_file_name*

StmEd reads your circuit file, rather than a data file created by a simulation, looking for waveform descriptions. Each of the input waveforms in the circuit file are available for inspection, and modification. StmEd's best feature may be that it not only allows you to view/modify these waveforms before simulating but allows you to see all of the waveforms together to check their synchronization.

StmEd uses the same configuration file (PSPICE.DEV) as Probe uses. As with Probe, you can "follow your nose" and discover the appropriate sequence of menus and commands to accomplish most operations. You may select menu items by hitting the key that matches the capital letter in the item (usually the first letter); or you may move the reverse video select region to the item, with the cursor keys, and then hit Enter; or you may click on the item with the mouse cursor. If you use a mouse, it will need to be a Microsoft-compatible mouse.

Your mouse may also be used for entering the vertices of piecewise linear (PWL) waveforms, directly.

When you are finished, StmEd writes out your circuit file including the textual information representing the edits you have made to the input waveforms.

D.4 REFERENCES: HOW SPICE WORKS

The details of the concepts and algorithms of circuit simulation, especially as they relate to SPICE and PSpice, are contained in the thesis:

L. W. NAGEL, *SPICE2: A Computer Program to Simulate Semiconductor Circuits*, Memorandum No. M520 (May 1975).

This thesis reviews and develops many of the methods that could be used for numerically simulating electronic circuits, and covers the advantages (and pitfalls) of these in great detail. The thesis is available for US$30 by sending a check payable to *The Regents of the University of California* to this address:

Cindy Manly-Fields
EECS/ERL Industrial Support Office
497 Cory Hall
University of California
Berkeley, CA 94720

Also interesting is a recently published book reviewing these algorithms and their use in a variety of simulators:

W. J. MCCALLA, *Fundamentals of Computer-Aided Circuit Simulation*, Kluwer Academic, 1988.

Devices

B, GaAsFET, 192-95, 247-49
C, capacitor, 249
D, diode, 184-90, 250-51
E, voltage-controlled voltage, 251-52
F, current-controlled current, 252
G, voltage-controlled current, 252-53
H, current-controlled voltage, 253-54
I, independent current source, 254
J, JFET, 190-92, 254-55
K, mutual coupling, 256
L, inductor, 257
M, MOSFET, 202-6, 258-63
N, digital input, 263
O, digital output, 263
Q, bipolar transistor, 195-202, 264-266
R, resistor, 267
S, voltage-controlled switch, 268
T, transmission line, 269-70
U, digital device, 270-71
V, independent voltage source, 271
W, current-controlled switch, 272
X, subcircuit call, 35-36, 273

Statements

.AC, small-signal analysis, 49, 218
.DC, DC sweep, 23-24, 218-19
.DISTO, SPICE2 distortion analysis, 133-34, 136
.DISTRIBUTION, user-defined distribution, 161, 220, 228
.END, end of circuit, 2, 220
.ENDS, end of subcircuit definition, 35, 221, 239

.EXTERNAL, external port, 221
.FOUR, Fourier analysis, 134-37, 221
.FUNC, function definition, 11, 222
.IC, initial bias-point condition, 126-27, 222, 229, 233
.INC, include file, 223
.LIB, library file, 36, 223
.LOADBIAS, load bias-point file, 224
.MC, Monte Carlo analysis, 163-64, 224-26
.MODEL, model definition, 147-48, 183, 223, 226-28
.NODESET, 127, 222, 229, 233
.NOISE, noise analysis, 102-3, 229-30
.OP, bias point, 7, 230
.OPTIONS, 4-5, 127, 152, 231-33, 241
.PARAM, parameter definition, 10, 109, 233-34
.PLOT, 24-26, 30, 49-50, 234
.PRINT, 24-26, 49-50, 234
.PROBE, 30-31, 50, 235
.SAVEBIAS, save bias-point file, 224, 235-36
.SENS, sensitivity analysis, 17-18, 237
.STEP, parametric analysis, 237-38
.STIMLIB, stimulus library file, 239
.STIMULUS, stimulus definition, 239
.SUBCKT, subcircuit definition, 35-36, 109-10, 223, 239-40
.TEMP, temperature, 151-52, 241
.TEXT, text parameter definition, 241
.TF, transfer function, 39, 43, 242
.TRAN, transient analysis, 122, 242-43
.WATCH, watch analysis results, 243
.WCASE, sensitivity and worst case analysis, 176-78, 244-45
.WIDTH, width of output, 232, 246

A

ABCD (chain) parameters, 75
ABSTOL, 127-28
AC analysis, *See* Frequency response
.AC (small-signal analysis) statement, 49, 218
Accuracy of transient analysis, 127
Active devices, 183-210
   bipolar junction transistor (BJT), 195-202
   diode, 184-90
   gallium-arsenide MESFET (GaAsFET), 192-95
   junction field-effect transistor (JFET), 190-92
   MOS field-effect transistor (MOSFET), 202-6
   nonlinear magnetics, 206-9
Amplitude modulation, 132
Analog Behavioral Modeling, 251-53

B

B (GaAsFET) device, 192-95, 247-49
B-H curves, 206-9
Backward ABCD parameters, 75
Bandlimited (pink) noise, 111-14
Base-width modulation, 197
Bias-point, DC, 4, 7, 230
Bipolar junction transistor (BJT), 195-202
   Q device, 264-266
Bode, H. W., 92
Bode plots, 58, 92-95, 98-99
Breakdown, reverse, 186

C

C (capacitor) device, 249
Capacitance, 149
   diffusion, 188-89
   diode, 188-90
   Miller, 212
   of network versus voltage, measuring, 129-30
   Raytheon and Curtice models, 193-95
   reverse-bias, 188-89
Capacitors, 7
   C device, 249
   in DC circuits, 14-15
   linear, 26-28
   scaling component values, 149-50
Chain (ABCD) parameters, 75
CHGTOL, 127-28
Closed-loop response:
   in inverse polar form, 97-98
   in polar form, 95-96
Comment (*) line, 3, 217
Complex values, 55-57
Component names, 8-9
Component values, 5-6
   scaling, 149-50
   sweeping, 151
Conductance, 27
Control shell, using, 275-76
Control statements, 217-46
Controlled sources:
   to "insert" noise, 107-9
   linear, 26-28
   multiple-input, 31-32
   polynomial, 28-29
Cross-modulation distortion, 133
Crossover distortion, 133
Current-controlled switch, W device, 272
Current gain, forward, 184, 198-99
Current loop gain, 83